Environmental Radiochemical Analysis III

Environmental Radiochemical Analysis III

Edited by

Peter Warwick
Department of Chemistry, Loughborough University, Loughborough, UK

RSC Publishing

The proceedings of the 10th International Symposium on Environmental Radiochemical Analysis held September 2006 in Oxford, UK.

Special Publication No. 312

ISBN: 978-0-85404-263-0

A catalogue record for this book is available from the British Library

Published by The Royal Society of Chemistry,
Thomas Graham House, Science Park, Milton Road,
Cambridge CB4 0WF, UK

Registered Charity Number 207890

For further information see our web site at www.rsc.org

Preface

The 10th Environmental Radiochemical Analysis Symposium of the Royal Society of Chemistry (RSC) was held in September 2006 at the Randolph Hotel in Oxford; at a time of renewed interest in radiochemistry in the UK and internationally. With a prospect of new nuclear build, a vigorous and targeted UK decommissioning programme and signs of increasing nuclear sector funding in UK universities the symposium benefited from great breadth and quality of papers, many from younger researchers. An abundance of overseas papers gave the symposium a truly cosmopolitan appearance and produced a very healthy forum for learning and cross-fertilisation of ideas.

This publication is the third in the successful series of RSC books of the proceedings of the symposium. Thanks are due to the editor, the secretary and the technical committee for the many hours they devoted to ensuring the quality of these papers.

My sincere thanks go to the Organising Committee:

Dr Anthony Ware, Treasurer, Environmental Energy Consultancy
Dr Kinson S Leonard, CEFAS
Professor Peter Warwick, Loughborough University
Dr Katherine Morris, University of Leeds
Dr Paul Martin, IAEA, Siebersdorf, Austria
Professor Syed Qaim, Jeulich, Germany
Dr Franz Schoenhofer, Vienna, Austria
Professor Bill Burnett, Florida State University, USA
Mr Mike Fern, Eichrom, USA

Special thanks to Mrs Carolyn Pickering, secretary of the Organising Committee

Dr Jon Martin
Chair of the Organising and Technical Committee of ERA10

Contents

RADIONUCLIDE ACCUMULATION AT A HYDROELECTRIC POWER DAM

E. Holm[1,2], Y. Ranebo[1,3], M. Eriksson[4], P. Roos[5], M. Peterson[1]

[1]Department of Medical Radiation Physics, Lund University Hospital, 221 85 Lund, Sweden
[2]Centro Nacional de Aceleradores, Avda. Thomas Alva Edison S/N, 41092 Seville, Spain
[3]European Commission, Joint Research Center, Institute for Transuranium Elements, Hermann von Helmholz-Platz 1, 76344 Eggenstein-Leopoldshafen, Germany
[4] IAEA-MEL, 4 Quai Antoine 1er, MC 98000, Monaco
[5]Risoe Nat. Lab., Roskilde 4000, Denmark

1 INTRODUCTION

There are about 1200 hydroelectric power plants in Sweden and several thousands of other water reservoirs for other purposes. World wide there are about 45 000 large dams in the world, the vast majority of which were constructed after 1950 and in total there are several millions of smaller ponds. These dams produce several benefits including supply of irrigation water, hydropower generation, flood control, recreation, fishing and others.

One hundred and ninety of the hydro electric plants in Sweden have depths larger than 13 m and up to 125 m. The history of these plants is well known with respect to physical parameters, construction year, etc. Generally the time of construction is different for different plants along a river. The constructed dams, which regulate the water flow, might act as flocculation basins. Very high sedimentation rates have been reported, up to 20 cm per year.[1] High sedimentation rates will limit the time of operation in shallow dams and increase the potential hazard if the sediments are reintroduced to the environment.[2] World wide there has been a large number of dam failures and in Sweden there have been 8 cases of serious floods and 2 cases of dam failures. Remedial actions for dams have been studied.[3,4]

The retention of dissolved silicated (DSi) artificial dams in Sweden and Finland has been studied which show a reduction in the delivery to the coastal zone.[5,6,7]

It is well known that anthropogenic radionuclides such as radiocaesium and plutonium together with natural ^{210}Pb are accumulated in sediments and can be used for the dating/growth rate determination of the sediments. The flux of these radionuclides depends on factors, such as physical and chemical properties, biological factors etc,. Water dams along rivers will stop the water flow and might act as effective traps by sedimentation processes and accumulate material that otherwise would be transported to the sea.

We have studied the accumulation and vertical distribution of ^{210}Pb, ^{137}Cs and $^{239+240}$Pu in sediment cores from a dam (Granö hydroelectric power dam) situated in the Mörrum River at SE Sweden (Figure 1, Table 1). This dam was constructed in 1958, about the time when the first large scale world wide fallout from nuclear test occurred.

There are in principle two sources for anthropogenic radioactivity in rivers in Sweden,1954-58 (20%) and 1961-62 (80%), and the Chernobyl accident, 1986. These events are quite distinct in time and by using isotopic ratios and radionuclide ratios the two sources can be distinguished from each other. China and France also conducted nuclear tests during 1960 -1980.

The dam at Granö was also selected for logistic reasons and the activity levels were expected to be high enough to allow meaningful analysis of the experimental data. Some basic parameters are given in the table below. The river receives water from a large upper lake, (area 150 km^2, 139 m over the sea level) and the catchment area of the lake is 3150 km^2. The lake is shallow (mean depth 3 m) and the residence time of the water is rather short. The water flow is variable but the annual average is 26 m^3 s^{-1}. The mean residence time of the water is then only 6.6 months.

Table 1 *Hydroelectric power plants in the Mörrum River (length of river 175 km)*

Name	*Construction year*	*Hight of fall (m)*	*Production (GWh)*
Granö	1958	18.5	31.5
Hemsjö ö.	1907	15	19.8
Hemsjö n.	1917	11.5	12.4
Marieberg	1918	4.8	4.8

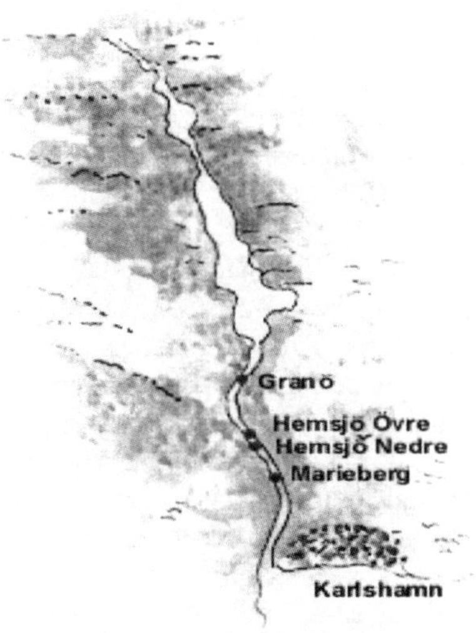

Figure 1 *Hydroelectric power dams along the Mörrum River.*

At the time of construction of the dam there will have been a decomposition of radionuclides (and other pollutants) trapped in vegetation, especially mosses and peat. Depending on depth, ventilation, etc. anoxic areas of the dam might develop. Granö hydroelectric power plant at Mörrumsån was constructed in 1958, i.e. just at the beginning of the nuclear area. This means that at that time only small amounts of radioactivity were trapped in biota in the flooded area.

2. METHODS AND RESULTS

2.1 Sampling and measurements

Sediment cores were taken in the accumulation area of the dams during the winter of 2004-2005 and the summer of 2005. A small boat was be used and during the summer and during winter time the ice was used as a sampling platform. The sediments were cores with a diameter of 31.16 or 55.39 cm. and were sliced in 1 cm. The maximum possible core depth was generally around 35 cm corresponding to the time of construction.

Radiocaesium (^{137}Cs, $T_{1/2}$ = 30 years) in the samples was measured using gamma spectrometry by solid state well-type HpGe gamma spectrometry.

Thereafter an aliquot of each slice is analysed for plutonium isotopes and ^{210}Pb. Plutonium was separated by anion exchange using ^{242}Pu as radiochemical yield determinant and measured by alpha spectrometry using solid state ion implanted Si detectors.

Pb-210 ($T_{1/2}$ = 21 years) was analysed after radiochemical separation of its daughter product ^{210}Po ($T_{1/2}$ = 163 days) and alpha spectrometry. ^{209}Po is used as radiochemical yield determinant.

2.2 Deposition and accumulation of ^{210}Pb.

Pb-210 was analysed in all cores. Typical distributions with depth are shown in Figures 2-7. The maximum possible coring depth was about 25-35 cm which should represent the time of construction of the dam. This corresponds to a sedimentation rate of 5-6 mm per year. The distribution in all cores is quite similar. It is evident that the sedimentation rate has not been constant. We have an abrupt discontinuity in the sediment record and dating by ^{120}Pb is not easy[8]. One can distinguish two periods. We see an early phase after the construction when the sedimentation rate was low, around 1 mm per year about 25-35 cm depth and then the rate increased to 20-30 mm per year. This rather sudden change can have several explanations. 40-50 years ago there was a significant ditching of the catchment area, especially peat bogs, to obtain more useable land for agriculture or foresting. Forest management techniques have changed such that clearance of larger areas is now commonplace, rather than harvesting of individual trees and both processes cause erosion, loss of vegetation and soil to the lake.

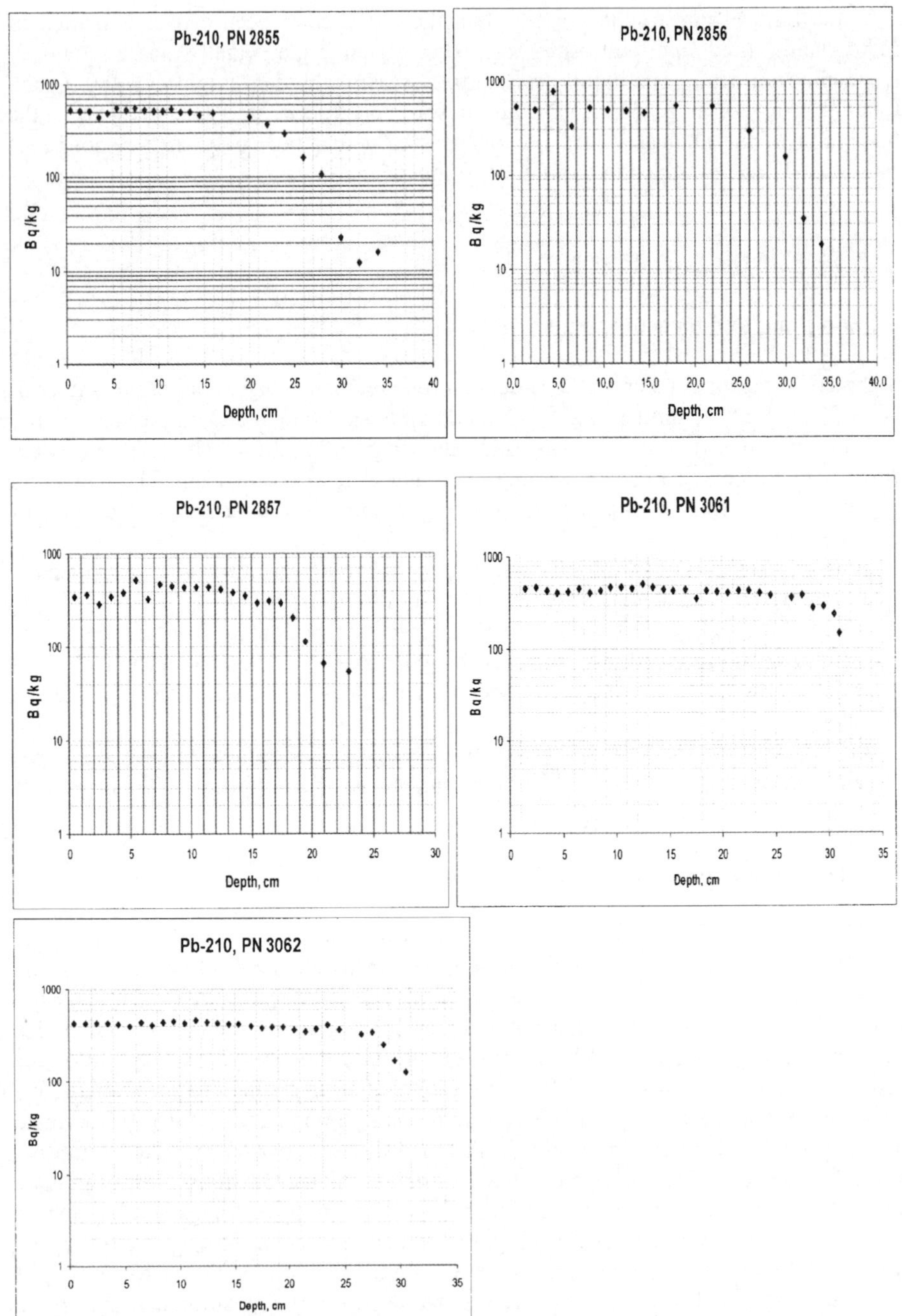

Figures 2-6. *Distribution with depth of ^{210}Pb in sediment cores.*

The area has been forested especially with pine and spruce and there is much less of agricultural land today. The humic material from the peat bogs will cause increased humic concentrations in the lake. Radionuclides can be associated with humic colloids and even radionuclides such as plutonium may be transported long distances.

The accumulated activity of ^{210}Pb from 0 to 35 cm varied from 11 000 -19 000 Bq m^{-2} with a mean of 14 200 +/- 3 400 Bq m^{-2}. We can not assume that ^{210}Pb has reached a state in the cores where decay equals the input to the sediments. This would require a core corresponding to 200 years. Anyway the average annual deposition has been larger than 360 – 600 Bq m^{-2} year^{-1} which is significantly higher than the expected annual flux for the area, about 70 Bq m^{-2} per year, while disturbed lakes in Sweden show fluxes 2-3 times higher than atmospheric ones.[9] The annual deposition today is around 200 Bq m^{-2} year^{-1}, but has been over 1000 Bq m^{-2} year^{-1} right after the change in sedimentation rate.

2.3 Accumulation of ^{137}Cs

The distributions of ^{137}Cs as a function of depth in the sediment cores are displayed in Figures 7-12. As can be seen a maximal concentration is observed at a depth of about 20 -24 cm with one exception where the maximum occurs at 10 cm (core 2857). In this core the maximal coring depth was also shorter-about 25 cm. This depth should for this core represent the year 1958 while for the other cores this year refers to 30-35 cm depth. If the depth for the maximal concentration represents 1963 then the sedimentation rate during 1958 to 1963 should have been about 3 cm per year which is in disagreement with the data for ^{210}Pb. The peaks are however broad. It is expected that there is some delay from initial deposition to maxima in the sediments due to the time needed for sedimentation processes. The input from the catchment area is in the early phase of fallout important especially during the Spring thaw. In our case we also have the change of the catchment area described under the accumulation of ^{210}Pb. These factors certainly would broaden the peak. We have then not even considered migration and diffusion in the sediments. If we look at the total area content (Bq m^{-2}) the maximal accumulation occurred at 30-32 cm except for core 2857 when it occurred at 21 cm. This is explained by the fact that the density of the sediments have a very low density in the upper part and much higher density deeper down. The accumulation actually started soon after the construction of the dam and using the integrated deposition as a basis for extrapolation we get a completely different sedimentation rate of about 2 mm per year which is in better agreement with the ^{210}Pb data. We must also consider the nuclear tests 1957-58, when 25% of the fallout took place.

The total deposition from nuclear test fallout was 2 300 Bq m^{-2} which today would have decayed to 870 Bq m^{-2}. The deposition from the Chernobyl accident was 1 400 Bq m^{-2} which today would have decayed to 900 Bq m^{2}. In total we then have about 1 770 Bq $^{-2}$. In the sediments we find a deposition between 2700 and 3 600 Bq m^{-2} with a mean of 3 400 +/- 950 Bq m^{-2}. The integrated deposition in the dam is about twice that from general integrated fallout. It is very common that the accumulation in the aquatic environment is higher than that from estimated fallout due to run off. The conclusion is that there is not any strong accumulation of radiocaesium in the dam. The annual accumulation today is in the order of 20-50 Bq m^{-2} year^{-1} compared to 700-800 Bq m^{-2} year^{-1} during the maximal fallout period. It shows that there is still some continuous transport of contaminated material from the catchment area and the lake downstream. The lake is very shallow and the sediments are redistributed by waves.

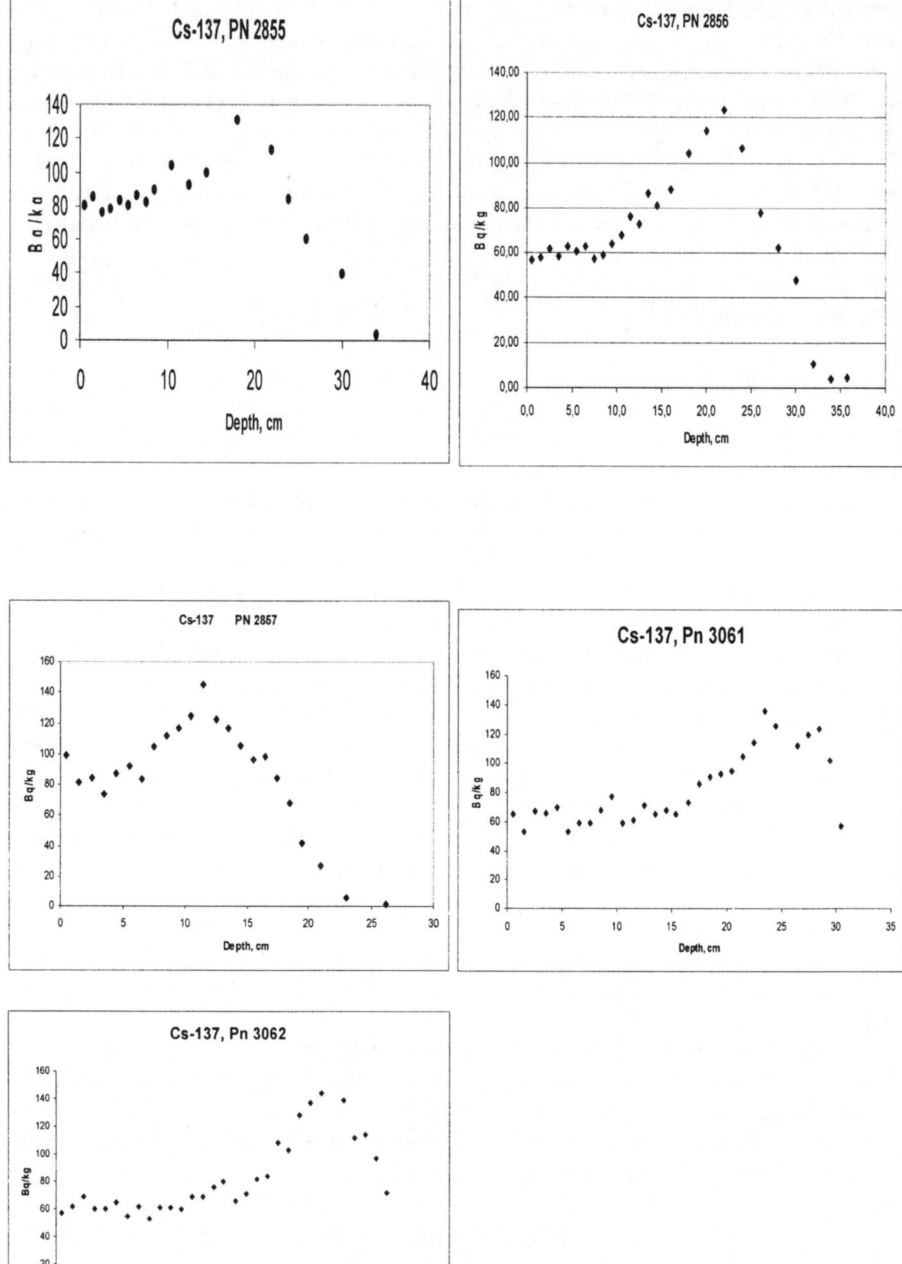

Figures 7-12 *Activity concentration of* ^{137}Cs *as a function of depth in the sediment cores.*

2.4 Accumulation of plutonium

The cores analyzed for $^{239+240}$Pu show a similar distribution as those for radiocaesium, i.e. maxima between 20-25 cm (Figure 13-14). In the same way by looking at annual deposition over time the accumulation started very soon after the construction of the dam. The total integrated deposition of $^{239+240}$Pu in the area from nuclear test fallout is about 27 Bq m^{-2}. The contribution from Chernobyl fallout was very small 1-2 Bq m^{-2}. In the sediment cores we find 170-330 Bq m^{-2} which is significantly higher than the general integrated fallout. The conclusion is that plutonium is significantly accumulated in the dam.

The present annual accumulation of plutonium in the dam is 0.6-1 Bq m^{-2} and was 30-90 Bq m^{-2} during the maximal fallout period.

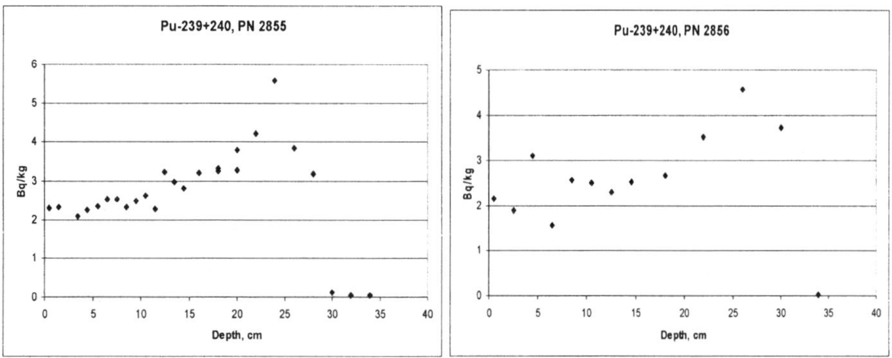

Figures 13-14. *Activity concentrations of $^{239+240}$Pu as a function of depth in two sediment cores.*

If we regard the activity ratio $^{239+240}$Pu/^{137}Cs as displayed in Figure 15-16 it will tell us something about the different behavior in the system between radiocaesium and plutonium. In fallout from nuclear tests the ratio was 0.012[9] and should today have increased to 0.031. Including the Chernobyl fallout the ratio should today be 0.015.[10] The accumulation of plutonium is substantially higher than that for radiocaesium during the whole period since 1958. This is the case especially during the nuclear test fallout period when this ratio is 0.07-0.08.

Generally plutonium is scavenged from the water column to a larger extent than caesium is. The ratio in open ocean water is 0.001 while in shallow areas, with high suspended load, both can be more or less completely removed to the sediments.

The sediments are very organic and anoxic conditions have been formed. It is known that plutonium forms organic humic complexes. In watersheds from the catchment areas the plutonium concentrations are as high as up to 300-400 μBq l^{-1} and ^{137}Cs, 30 mBq l^{-1}. In the major lake from where the river starts the concentrations are respectively 100 and 10 times lower. Our data suggests that concentrations downstream is due to bunding and transport by organic sediments which are transported by the river and trapped in the dam.

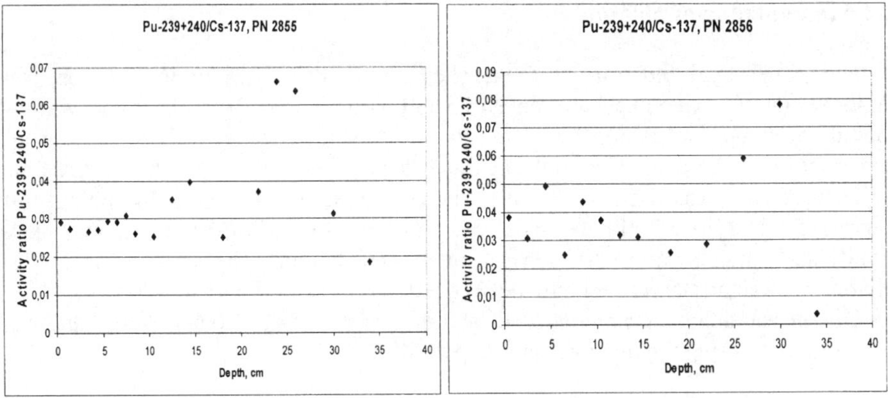

Figures 15-16 *The activity ratio* $^{239+240}Pu/^{137}Cs$ *as a function of depth in two sediment cores.*

3 CONCLUSION

The use of radio lead (^{210}Pb) for dating of the sediments in a dam constructed for hydroelectric power showed a dynamic pattern. The sedimentation rate based on the ^{210}Pb data was low after the construction of the dam, 1 mm per year and increased then to 20-30 mm per year. The reason for this increase in sedimentation rate is probably due to ditching of the catchment area and that forest management techniques have changed such that clearance of larger areas is now commonplace, rather than harvesting of individual trees and both processes cause erosion, loss of vegetation and soil to the lake.
changes in logging of forest and use of land bringing large amounts of material into the upper lake and further down the river system.

The average annual deposition has been larger than 360 – 600 Bq m^{-2} $year^{-1}$ which is much higher than the expected annual flux for the area, about 150 Bq m^{-2} $year^{-1}$. The maximal concentrations of the anthropogenic radionuclides ^{137}Cs and $^{239+240}Pu$ occurred at a depth of about 20 cm which should then correspond to the major nuclear tests around 1962-63. Radiocaesium shows a different accumulation pattern and/or higher mobility in the sediments than plutonium and the 1963-peak is quite broad. The radiocaesium fallout from the Chernobyl accident was small in this area and the peak from this event is either rather small or overlaps the distribution from nuclear test fallout, which also broadens the peak compared to plutonium. Radiocaesium is not to any higher degree accumulated in the dam compared to other fresh water systems. Applying a time resolution, Full Width of Half Maximum, gives 10-12 years for ^{137}Cs while for $^{239+240}Pu$ this time resolution is about 6 years.

The accumulation of ^{210}Pb and plutonium is up to 10 times higher than the general integrated deposition in the area.

References

1 B.L.Valero-Garces, A. Navas, J. Machín, D. Walling, 1999, *Geomorphology*, **28**, 23

2 S.J.Bennett, C.M. Cooper, J.C. Ritchie, J.A. Dunbar, P.M. Allen, L.W. Caldwell, T.M. McGee, 2002, *J. of the American Water Resources Association*, **38**, 1307.

3 R.J.Wasson, G. Caitcheon, A.S. Murray, M. McCalloch, J. Quade, 2001, *Environment Management*, **29**, 634.

4 A. Palmieri, F. Shah, A. Dinar, 2001, *J. Environm. Management*, **61**, 149.

5 D.J Conley, P. Stålnacke, H. Pitkänen, A. Wilander, 2000, *Limnology and Oceanography*, **45**, 1850.

6 C. Humborg, D.J. Conley, L. Rahm, F. Wulff, A. Cosinasu, V. Ittekkot, 2000. *AMBIO*, **29**, 45

7 C. Humborg, S. Blomqvist, E. Avesan, Y. Bergensund, E. Smedberg, 2002 *Global Biogeochemical Cycles*, **16**. Art No. 1039.

8 P.G. Appleby, 1998, NKS, Seminar, Helsinki 2-3 April, 1997, *STUK A 145*, 6

9 F. El-Daoushy, *Environment International*, 1988,**14**, 305

10 E. Holm, B. Persson, 1975, *Health Physics*, **29**, 43

DETERMINATION OF THE TRANSFER OF TRITIUM TO CROPS FERTILISED WITH CONTAMINATED SEWAGE SLUDGE

G J Ham, B T Wilkins and D Wilding

Radiation Protection Division, Health Protection Agency, Chilton, Didcot, Oxfordshire, OX11 0RQ, UK

1 INTRODUCTION

Sewage sludge is being increasingly used as a soil conditioner. Authorised discharges of radionuclides into sewers can result in contamination of the sludge, and so provide a pathway by which radionuclides can enter the foodchain. A review of published information indicated that specific data on the uptake of tritium from sludge amended land into crops were scarce[1]. Discharges from a radiopharmaceutical plant in Cardiff are presently routed to a wastewater treatment works, and have resulted in elevated concentrations of tritium in sewage sludge pellets. This provided the potential to generate specific data on tritium uptake by crops grown in amended soil.

This research opportunity is being exploited at the Chilton laboratory of the Health Protection Agency's Radiological Protection Division (RPD), the work being funded by the Food Standards Agency (FSA). The aim is to make use of the RPD's existing lysimeter facility to generate realistic uptake data for selected crops grown in three soil types, each amended with tritium contaminated sludge. This is to be augmented by smaller scale studies on a wider range of crops that would generate relative values for uptake parameters, based on the same soils. A combination of these two sets of data should then provide parameter values that can be applied in predictive models.

There is considerable uncertainty about the rate at which tritium can be lost from soil, either via transpiration or via downward migration. This has important implications for situations involving the application of tritium-containing sludge to agricultural land because of the constraints of the Sludge (Use in Agriculture) Regulations (commonly referred to as the Safe Sludge Matrix). These regulations require that crops cannot be harvested from treated land until at least 10 months after the sludge has been applied. The aims of the first phase of this project, carried out during 2005, were to determine whether uptake was measurable in crops grown in treated soil and to investigate whether tritium was being lost rapidly from the soil. It was recognised that, in terms of crop production, this part of the study was outside the constraints of the Safe Sludge Matrix. In this paper the results from 2005 are used to derive implications for other studies investigating the transfer of tritium from soil to plant, and for the second phase of the this project that is taking place in 2006.

2 STUDY DESIGN AND INITIATION

2.1 Starting materials

About 300 kg of sewage sludge pellets were collected from the Cardiff East sewage works in late 2004. An initial analysis of the tritium content showed an average concentration of 87 kBq kg^{-1} of total tritium. Nearly all of the activity was in an organically bound form.

The lysimeters at RPD are 1.7 m diameter and 0.5 m deep and set out in three pairs, with each pair containing a different soil type. The soil types employed are Hamble loam, Fifield sand and Adventurers peat. Each soil had been collected from suitable field sites in 1983. They had not been previously amended with tritium.

In order to allow the study of a larger range of crops and soil types the existing lysimeters were supplemented by a series of smaller containers. Water cisterns of 20 l capacity were used with dimensions of 0.45m * 0.3m * 0.27m (w*d*h). To provide the soils for these new containers, further samples were collected during the early spring of 2005 from the locations originally used when setting up the lysimeters. Samples of Denchworth clay were also collected from a site that had been identified in 1983. Together, these four soils provide a suitable contrast in the types of agricultural soil found across the UK. Sets of 18 containers were filled with each type of soil, tilled and allowed to settle. The soils in the lysimeters were dug over and allowed to settle at about the same time.

2.2 Study design

The investigation of soil to plant transfer of tritium presents unique problems because of possible losses from the soil by transpiration as well as by leaching and plant uptake. Not only is it very difficult to measure this transpired activity but there is also the potential to transfer activity to neighbouring containers and plants. Plants may also be affected by atmospheric tritium. These factors made the use of control crops an imperative.

One important objective for 2005 was to improve understanding of the behaviour of tritium in soil. The lysimeters had not been designed to enable the collection of leachate. Consequently, for each soil type, one of the smaller containers was modified for this purpose. The modified containers were planted with grass, because after amendment and sowing, the soil would not be disturbed. The RPD has no facility to measure tritium transpired from the soil. It was therefore decided to take samples of surface soil from the lysimeters throughout the growing season in order to observe any changes in tritium concentration with time. Because of the containers' smaller surface area it was not possible to take repeated soil samples without disturbing the crops. After collection, all soil samples were sieved through an approximately 3 mm mesh, an aliquot was dried at 105^{0}C to determine moisture content and a second aliquot stored at -20^{0}C for tritium analysis.

The crops chosen were dwarf french (DF) beans (the Prince), cabbage (Golden Acre), carrot (Early Nantes), grass (Twystar Lollium), strawberries (Korona) and potatoes (Wilja). The crops were grown so that each soil : crop combination involved 3 of the smaller containers. Two of these were treated with sludge pellets and the third used as a control.

Once the soil had settled after tillage, the designated containers and the six lysimeters were amended with sewage sludge pellets at a rate corresponding to about 20 tonnes per hectare. This exceeds the maximum rate specified in the Safe Sludge Matrix, but was considered necessary because the measured concentrations of tritium in the pellets were lower than had originally been expected. The pellets were incorporated into the top

100 mm of soil using standard gardening tools. The soils were mixed again 3 and 7 days after amendment.

All six crops were sown or planted about 1-2 weeks after contamination in late May 2005 in individual small containers. In addition, the lysimeters were planted with grass, potatoes and carrots. They were maintained according to good horticultural practice, watered as required and regularly fed with a proprietary liquid feed. Grass was cut to about 30 mm above the soil surface whenever it reached about 150 mm in height. All other crops were collected at maturity. All crops except grass were washed in tap water and inedible parts discarded. After roughly chopping or shredding the entire sample, an aliquot was dried at 105^0C to determine moisture content and a second aliquot stored at –20^0C for tritium analysis. Potatoes and carrots were not peeled but for one potato crop grown in a lysimeter an additional sample was taken for separate analyses of peel and flesh.

2.3 Analyses

The determination of tritium in solid environmental materials was based on controlled combustion followed by liquid scintillation counting using a low-level Quantalus instrument that had been suitably calibrated.

2.3.1 Procedure for solid samples. All crop and soil samples were analysed by combustion in a Raddec® pyrolyser. The pyrolyser first removes water from the sample by slowly raising the temperature to 180^0C and maintaining it at that temperature for 45 minutes. A stream of air carries the water vapour over a Pt-alumina catalyst at 800^0C and then into a trap solution of dilute nitric acid. At the end of this cycle the trap solution is replaced with fresh nitric acid solution and the sample temperature slowly raised to 550^0C. The air is then replaced by oxygen and the sample is maintained at that temperature in the oxygen atmosphere for at least 12 hours. In all cases, approximately 10 g aliquots of crop and 5 g aliquots of soil were analysed.
The trap solution from the lower temperature is defined as containing the aqueous tritium from the sample and that from the higher temperature the organically bound tritium (OBT).

2.3.2 Procedure for liquid samples. Rainwater and leachate samples were prepared by distillation. Where total tritium was required potassium permanganate was added and the sample boiled under a reflux condenser for 30 minutes before distilling an aliquot for measurement.

2.3.3 Liquid scintillation counting. An 8 ml aliquot of the trap or distilled solution is mixed with 12 ml of Meridian Goldstar® scintillant and counted in a Quantulus® liquid scintillation counter against a reagent blank.

2.3.4 Results. All results were calculated as Bq g^{-1} with respect to fresh mass for crops and "as collected" mass for soils. All individual results are reported with their uncertainties based on standard analytical uncertainties multiplied by a coverage factor of k=2 which provides a level of confidence of approximately 95% confidence interval. However, no allowance has been made for the uncertainty arising from sampling.

Table 1 *Total tritium concentration in potatoes grown in lysimeters, Bq g^{-1} fresh mass.*

	Loam soil	*Peat soil*	*Sandy soil*
Measurement 1	0.143 ± .0020	0.080 ± 0.014	0.225 ± 0.026
Measurement 2	0.148 ± 0.021	0.102 ± 0.017	0.226 ± 0.030
Measurement 3	0.135 ± 0.021		

2.4 Validation of analytical method

2.4.1 Pyrolyser. The catalyst was changed after analysing 15 samples. The recovery efficiency was checked by analysing a known mass of calibrated tritiated thymidine carried on filter paper. This was performed at the beginning and end of the life of the catalyst. After about a year's experience we have recorded a mean efficiency of 90 ± 10%. After each sample run a blank sample of moistened filter paper was analysed to confirm that no cross contamination was occurring between samples.

The manufacturer recommends that the sample be kept at 500^{0}C under oxygen for one hour to complete combustion. However, when these recommendations were followed, tritium was measured in those blank runs that followed the analysis of sludge pellets or amended soil. Extending the burn time to 90 minutes and increasing the temperature to 550^{0}C, reduced but did not eliminate this carry over. Extending the final burn period to more than 90 minutes would have made it impossible to complete a run within a working day. Finally it was decided to continue the final burn overnight. This approach eliminated the problem.

2.4.2 Quantulus liquid scintillation counter. Counting efficiency standards were prepared as described in paragraph 2.3.3 using 8 ml of distilled water with a weighed aliquot of about 0.1g of standardised aqueous tritium. Each sample batch was counted with a standard and blank. The measured counting efficiency and background were checked for consistency with established means and used to calculate the results for that batch. The mean counting efficiency was 24.5 ± 0.5% and background was 1.6 ± 0.8 CPM.

2.4.3 Analytical reproducibility. This was confirmed by measurements made on samples of potato grown as part of the 2005 study. Activity concentrations in potato flesh should be uniform. The results are shown in Table 1. There was no significant difference between replicate measurements, which strongly supports the reproducibility of the analytical procedure.

2.4.4 Overall validation. The analytical procedure has been tested recently via participation in intercomparison exercises. One organised by the National Physical Laboratory involving tritiated water and another involving urine organised by BfS Germany. Both samples were prepared both by distillation and by pyrolysation. In all cases, the results from HPA-RPD were in good agreement with the expected values.

3 RESULTS AND DISCUSSION

3.1 Concentrations of tritium in soil

The results for the total tritium concentration in sandy soil from the lysimeters are shown in Figure 1. A similar pattern was found for all soil types in both lysimeters and tubs.

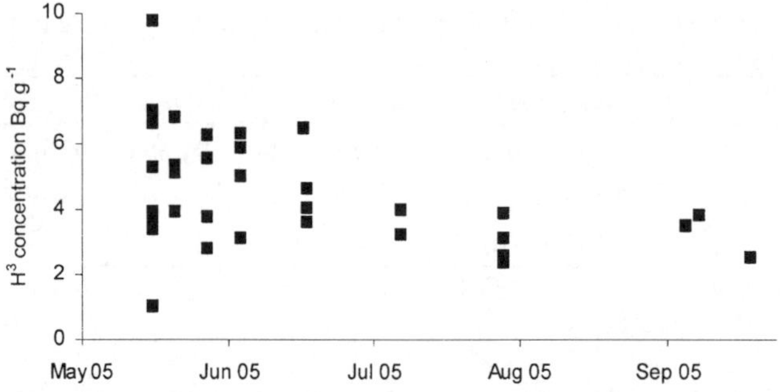

Lysimeter: tritium concentration in sandy soil

These indicate a large variability in the individual results, which tended to decrease with time. The reproducibility of the analytical method has been established. Consequently, the variability can probably be ascribed to sample inhomogeneity.

In the early stages following the amendment of the soil, individual sludge pellets were still visible in the sample and some passed through the sieve intact. This would obviously lead to homogeneity problems. The physical break up of pellets during the growing season would be consistent with the reduced variability over time.

In Table 2, the average concentrations for all samples taken in May and June are compared with those taken in September. Because of both the high variability between individual results and the different number of values making up each average, the uncertainties on these values are difficult to estimate but are about 50% on the May / June values and 25% on the September ones.

Table 2 *Change in tritium concentration in surface soil through the 2005 season.**

Soil type and Period		*Average concentration for the period, Bq g^{-1}* Aqueous	*Organically bound*	*Total*	*% decrease in the growing season*
Lysimeters					
Loam	May / June	0.14	2.47	2.61	
	September	<0.02	1.07	1.09	58%
Peat	May / June	0.29	4.70	4.99	
	September	0.06	3.25	3.31	34%
Sand	May / June	0.12	2.09	2.20	
	September	<0.02	1.01	1.03	53%
Containers					
Loam	May	0.27	3.28	3.55	
	August / September	0.02	1.21	1.23	65%
Peat	May	0.34	4.04	4.38	
	August / September	0.05	3.17	3.22	27%
Sand	May	0.16	3.23	3.39	
	August / September	0.02	1.11	1.12	67%
Clay	May	0.36	3.86	4.22	
	August / September	0.02	1.38	1.40	67%

*Uncertainties are difficult to calculate, but are between 25 and 50%, see text.

Table 3 *Variation in concentrations of tritium with depth in lysimeters on 25/8/05*

Depth, mm	*Measured Tritium, Bq g^{-1}*					
	Loam		*Peat*		*Sandy*	
	Aqueous	*Organically Bound*	*Aqueous*	*Organically Bound*	*Aqueous*	*Organically Bound*
0-100	0.04 ± 0.02	0.67 ± 0.07	0.03 ± 0.02	1.2 ± 0.1	<0.02	0.70 ± 0.08
100 - 200	0.02 ± 0.01	0.10 ± 0.02	<0.02	0.35 ± 0.04	<0.02	0.04 ± 0.01
200 - 300	0.03 ± 0.02	0.02 ± 0.01	0.03 ± 0.02	0.04 ± 0.02	<0.02	<0.02

Despite the variability, some broad trends can be discerned. In the lysimeters, the decrease in the concentration of tritium over the growing season was approximately 50% for both the loam and the sand but only just over 30% for the peat. The loss of tritium in the small containers followed a similar pattern. The apparent differences in the values given in Table 2 may be because of the large variability in individual results. For all of the soil types, tritium in the surface layer was predominantly present as OBT, the values for aqueous tritium at the end of the season being at or close to the detection limit.

The size of the lysimeters was sufficient for soil cores to be taken just before the harvesting of crops. Concentrations of tritiated water and OBT were determined at different depths in the soil. The results summarised in Table 3, showed that the organically bound tritium remained largely in the upper layer which represents the depth of tillage. The concentration of aqueous tritium was low and often below detection limits. However, where measurable values were obtained, concentrations were relatively uniform with depth.

3.2 Concentrations of tritium in leachate and rainwater

For each soil type, the leachate was collected from one of the treated small containers being used to grow grass. These were considered the most suitable because the underlying soil was not going to be disturbed after the crop had been sown. The volume of leachate was measured so that a total flux of tritium out of the container could be estimated. In most cases, analyses were confined to total tritium, but for a few samples the tritiated water content was also measured. The total tritium lost into leachate from the containers over the 2005 season is summarised in Table 4.

Table 4 *Tritium in container leachate May to November 2005*

Soil type	*Tritium in leachate Bq*	*Proportion of activity added to container.*	*Proportion of activity lost, derived from measurements in soil*
Loam	3,300 ± 300	25%	65%
Peat	1,900 ± 200	9%	27%
Sand	4,600 ± 400	21%	67%
Clay	5,100 ± 400	23%	67%

All soil types showed a similar pattern of tritium loss to leachate with time. After a short period of about a month, the tritium concentration in the leachate peaked at about 100 – 200 mBq ml^{-1}. It then fell over the next month or so to a low and fairly constant value of about 10 – 20 mBq ml^{-1}. Where measurements were made, there was no significant difference between the aqueous and total tritium concentration. It would therefore be reasonable to assume that all of the tritium lost via downward leaching was in the form of tritiated water. Although tritium could be measured in the leachate, the overall flux did not account for all of the losses as broadly implied from the measurements in soils from the containers. Table 4 shows that for all soil types, only about one third of the decrease in soil activity can be accounted for in the leachate.

It is also worth noting that the fluxes of tritiated water out of the small containers were lowest in the late summer at about the time that the soil cores were taken from the lysimeters (Table 3). This seems reasonably consistent with the data in Table 3, which indicated that most of the activity still in the soil was retained in the surface layer in the form of OBT.

The concentration of tritium in rainwater was measured throughout the season. Apart from one positive reading in August, all measurements were less than the detection limit of 0.006 Bq ml^{-1}. The collection of the anomalous sample coincided with the harvesting of both the carrots and potatoes during dry weather. The rainfall collector may then have been contaminated with wind-blown soil. Overall, the measurements indicate that any tritium in rainwater was not contributing significantly to the concentrations observed in the leachate.

3.3 Vegetables

The measured activity concentrations of tritium for crops grown in uncontaminated soil were generally very low, many being below the limit of detection. The tritium content of all crops grown in contaminated soil were measurable but low. Generally however, where analyses were carried out on the same crop and soil type in two separate containers, the results were not significantly different from each other. While the dominant form of tritium in soil was as OBT, the tritium in crops was nearly all present as tritiated water.

Table 5 *Average concentration of tritiated water in crops.*

Soil	*Experiment*	*Tritium concentration (Bq g^{-1})*			
		Cabbage	*Carrots*	*DF Beans*	*Potatoes*
Loam	Container	0.05 ± 0.01	0.016 ± 0.007	0.05 ± 0.02	0.04 ± 0.01
	Lysimeter		0.10 ± 0.02		0.12 ± 0.02
Peat	Container	0.025 ± 0.01	0.03 ± 0.01	0.05 ± 0.02	0.05 ± 0.01
	Lysimeter		0.12 ± 0.02		0.08 ± 0.01
Sandy	Container	0.05 ± 0.01	0.05 ± 0.02	0.05 ± 0.02	0.07 ± 0.01
	Lysimeter		0.11 ± 0.02		0.20 ± 0.02
Clay	Container	0.035 ± 0.01	0.025 ± 0.01	0.04 ± 0.02	0.04 ± 0.01

Table 5 gives the average measured concentrations of tritium in vegetables in terms of fresh mass. For the small containers, the variability in concentrations of tritiated water between the same crop grown in different soil types was fairly small and comparable with the variability between replicate containers. However, there was a significant difference between the concentrations of tritiated water in crops grown in the small containers compared with those for the same crop grown in lysimeters. In all cases, the values for crops grown in lysimeters were significantly greater than the corresponding values for the small containers, while there were no similar differences in the concentrations of tritium in soil. This supports the idea that the smaller containers cannot be used to provide absolute values for transfer parameters. It is not possible to estimate soil to crop transfer factors because of the changes in the concentration of tritium in soil over the growing season.

3.4 Grass

Measurements of the tritium content of grass samples cut throughout the season were very variable and between 30% to 80% of the total tritium was in an organically bound form. This is observation differed significantly from the other crops, were little or no organically bound tritium was found. Also, a number of the control grass samples contained measurable tritium concentrations while none was found in other control crops. Earlier studies have noted that grass can be contaminated by soil splash[2]. The presence of contaminated soil would explain both the variability and the high contribution from organically bound tritium.

4 CONCLUSIONS

Concentrations of tritium in soil decreased throughout the season. Measured values in the soil itself displayed considerable variability, although this improved with time. The trend in improved homogeneity with time might have been expected given the form of the activity applied and the relatively short time since application. Visual observations made during sample processing also supported this view. Tritium remaining in the soil was largely in the form of OBT, and in the lysimeters was retained in the surface (0 – 100 mm) layer.

The results of measurements in leachate from the small containers indicate a marked loss of tritium in the form of tritiated water. However, the total mount of activity lost was insufficient to account for losses implied from the measurements in soil.

Concentrations of tritium were measurable in all crops grown in contaminated soil. With the exception of grass, the activity was mainly in the form of tritiated water. For grass, the results indicated that soil splash was an important contributor to the activity observed in the vegetation. Again with the exception of grass, activity concentrations in all crops grown in control containers were near to, or below the limit of detection. The results for grass grown in the control containers indicated that some contamination with soil had occurred.

The reproducibility between individual soil : crop combinations in the small containers was good. However, the concentrations of tritium in crops grown in lysimeters were significantly greater than the corresponding values in the small containers by factors of between 2 and 5. The lysimeters are more likely to represent field conditions, consequently, the small containers cannot be used to provide absolute values for transfer parameters. No significant difference in tritium concentration was measured between different crops or soils.

5 IMPLICATIONS FOR OTHER STUDIES OF TRITIUM UPTAKE BY PLANTS

Since all of the tritium in the crops was in the form of tritiated water, it seems reasonable to assume that the uptake into the plants derives from the aqueous fraction in the soil. The bulk of the tritium in the rooting zone in the soil was however still in the form of OBT. The uptake into the crop would therefore appear to come from a mobile pool of tritiated water. This pool would be subject to losses via transfer into the plant, downward leaching out of the rooting zone and transpiration into the atmosphere as water vapour.

From the measurements of tritiated water in the leachate made so far in this study, a significant fraction of the activity originally applied had been converted to tritiated water and had migrated out of the rooting zone. However, the activity concentrations observed in crops grown in lysimeters were higher than the corresponding values for crops grown in smaller containers. This might suggest that the amount of activity leached downwards was less in the lysimeters. Taken together, these results indicate that the smaller containers cannot be used to provide uptake data that replicate what might be observed in field conditions. In this established lysimeter facility it is not possible to estimate the amount of activity transported downwards out of the rooting zone. However, such a capability would be desirable in any new facility intended for the study of tritium.

It is likely that transpiration would be greater in smaller containers, simply because the bulk soil would heat up more than would be the case in larger lysimeters. There is no evidence from the present study to confirm or refute this, but it is another reason for using larger scale experiments to try to replicate field conditions adequately. The rigorous estimation of losses via transpiration requires a more sophisticated apparatus that is not commonly available. It might be possible to estimate the importance of this process using an activity balance approach, although there could be large uncertainties in estimating the total activity transferred to the entire biomass. For example, in crops such as strawberries some leaves could have already decayed before any berries were ready for harvest. The total amount of activity transferred to the crops has not been estimated in the present study. However, given that the losses via leaching did not account for all of the changes in the amounts in surface soil, significant losses via transpiration cannot be ruled out.

The overall implications are that simple studies of the uptake of tritium from sludge-amended soil to crops are unlikely to provide transfer parameters in the conventional form used for many other radionuclides, ie the soil : plant transfer factor approach. The next phase of the present study will include further monitoring of losses via leachate in the smaller containers, but will focus on the uptake into crops grown in the lysimeters. Crops grown in 2006 will conform to the requirements of the Safe Sludge matrix, and so should provide an estimate of the broad radiological importance of amending agricultural land with tritium-containing sludge.

References

1 G. J. Ham, S. Shaw, G. M. Crockett and B. T. Wilkins, Partitoning of radionuclides with sewage sludge and transfer along the terrestrial foodchan pathways from sludge-amended land – A review of data. National Radiological Protection Board, NRPB-W32, 2003, ISBN 0 85951 504 4.

2 N. Green, D. Johnson and B. T. Wilkins, *J. Environ. Radioactivity,* **30** (2), 173

TECHNETIUM-99 (^{99}Tc) IN MARINE FOOD WEBS IN NORWEGIAN SEAS-RESULTS FROM THE NORWEGIAN RADNOR PROJECT

H. E. Heldal[1,2], K. Sjøtun[3] and J. P. Gwynn[4]

[1]Institute of Marine Research (IMR), PO Box 1870 Nordnes, N-5817 Bergen, Norway
[2]University of Bergen, Department of Chemistry, PO Box 7800, N-5020 Bergen, Norway
[3]University of Bergen, Department of Biology, PO Box 7800, N-5020 Bergen, Norway
[4]Norwegian Radiation Protection Authority (NRPA), Polar Environmental Centre, N-9296 Tromsø, Norway

1 INTRODUCTION

Due to oceanic long-range transport of authorized discharges from the reprocessing plant Sellafield in Cumbria (UK), Norwegian coastal areas have been exposed to substantial amounts of technetium-99 (^{99}Tc). Several research programs have been initiated in order to investigate the environmental impact of ^{99}Tc along the Norwegian coast. One such program is RADNOR (Radioactive dose assessment improvements for the Nordic marine environment: Transport and environmental impact of technetium-99 (^{99}Tc) in marine ecosystems). The RADNOR consortium is composed of the Norwegian Radiation Protection Authority (NRPA) (coordinator), the Norwegian University of Life Sciences (UMB) and the Institute of Marine Research (IMR). This paper presents results from a study carried out at IMR within the RADNOR program.

The aims of the study were 1) implementation of an analytical method for measuring ^{99}Tc in environmental samples, and 2) investigation of the uptake and accumulation of ^{99}Tc in marine food webs in Norwegian waters. In the present paper we focus on the uptake and accumulation of ^{99}Tc in benthic species.

The increases in the ^{99}Tc concentrations in seawater and some marine organisms in Norwegian waters have been subject to large public concern. A documentation of the contamination levels in fish and seafood is of great importance to the Norwegian fisheries' credibility towards both domestic and foreign markets. Results from this study will also form the basis for improved assessments of doses from seafood consumption.

2 MATERIALS AND METHODS

2.1 Sample collection

During 2003-2006, samples of wolf-fish (*Anarhichas* sp.), European lobster (*Homarus gammarus*), crab (*Cancer pagurus*), edible sea urchin (*Echinus* sp.), blue mussel (*Mytilus edulis*), snails (*Littorina* spp. and *Patella vulgata*), brown seaweeds (*Ascophyllum nodosum*, *Fucus vesiculosus* and *Fucus serratus*) and seawater were collected from five locations along the Norwegian coast (Arendal (Aust-Agder), Tysnes (Hordaland), Espegrend (Hordaland), Værlandet (Sogn og Fjordane) and Rørvik (Nord-Trøndelag)

(Figure 1). In addition, five samples of king crab (*Paralithodes camtschatica*) have been collected from the Varanger fjord in Finnmark County in 2003 and 2004. All samples were stored deep-frozen and transported to IMR where they were subsequently ground, freeze-dried and homogenized. 100 L of seawater were collected along with biota samples in order to be able to calculate CFs.

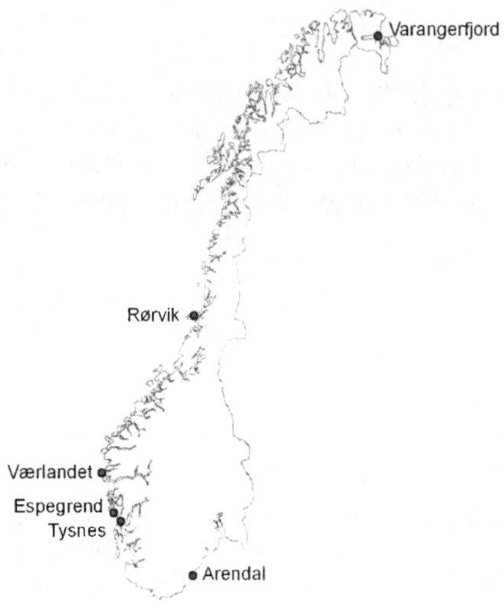

Figure 1 *Sampling locations on the coast of Norway*

2.2 Analytical method

^{99}Tc is a pure beta-emitter ($E_{\beta max}$=293keV) decaying from ^{99}Mo, and a radiochemical separation of this radionuclide is required for quantitative analysis. The detailed analytical method is described by Harvey *et al.*[1]. Briefly, rhenium is added in the form of $KReO_4$ as a yield monitor, and a preliminary extraction of ^{99}Tc (and Re) based on anion-exchange separation is performed. After iron hydroxide scavenging, ^{99}Tc and Re are further extracted by a second anion-exchange and subsequent sulphide precipitation. Finally, their tetraphenyl arsonium salts are isolated. ^{99}Tc is determined by beta-counting on a RISØ beta-counter, and the yield of the rhenium tetraphenyl arsonium salt is determined gravimetrically.

IMR has participated in one international intercalibration test, where the analytical results gave good comparisons with other methodologies. The method has a theoretical detection limit of 0.08 Bq/kg wet weight (w. w.), with a combined analytical uncertainty due to the counting statistics and chemical yield of generally less than 5%, based on a counting time of 48 hours.

3 RESULTS AND DISCUSSION

3.1 ^{99}Tc in marine organisms along the Norwegian coast

Levels of ^{99}Tc in seawater and biota increased significantly along the Norwegian coast after EARP began operation in 1994 (Brown *et al.*[2] Kershaw *et al.*[3]). The ^{99}Tc concentration in the Norwegian Coastal Current (NWCC) outside Tromsø in June/July

1999 was 1.49 Bq m^{-3} (Kershaw *et al.*[3].). This is about an order of magnitude higher than the concentration found at the same location by Kershaw *et al.*[4] in 1994 (0.14 Bq m^{-3}).

The ^{99}Tc concentration in the brown seaweed *Fucus vesiculosus* increased by up to five times in the outer Oslofjord from 1996 to 1997 (Brown *et al.*[2]). Elevated ^{99}Tc levels were also found in lobsters (*Homarus gammarus*) (≤ 42 Bq kg^{-1} w. w.) during the same time period (Brown *et al.*[2]). Conversely, the ^{99}Tc concentrations in e.g. mussels (*Mytilus edulis*) collected in the outer Oslofjord in 1997 were low (0.68 Bq kg^{-1} w. w.) (Brown *et al.*[2]).

In this study, we found, as expected, the highest ^{99}Tc concentrations in brown seaweeds (Table 1, Table 2). The ^{99}Tc concentrations reached 964 Bq/kg d. w. (dry weight) in an old part of the brown seaweed *Ascophyllum nodosum* (Table 2). Lobsters contained up to 42 Bq/kg w. w. (Table 1). Further, we found elevated levels in snails (*Littorina* sp.) living close to brown seaweeds. In all other biota samples measured, the ^{99}Tc concentrations were close to the detection limit, and not above 2.18 Bq/kg w. w..

Although the levels have increased, we conclude that currently, ^{99}Tc-levels in marine food webs in Norwegian waters are generally low and represent, to our knowledge, no threat to human health.

Table 1. *^{99}Tc (Bq/kg wet weight (w. w.)) in marine organisms along the Norwegian coast 2003-2006*

Species		n	^{99}Tc (Bq/kg w. w.)	
			min	max
European lobster (M)	*Homarus gammarus*	23	0.9	7.8
European lobster (F)	*Homarus gammarus*	22	3.6	42
Crab	*Cancer pagurus*	10	bd	2.18
King crab	*Paralithodes camtschatica*	3	bd	bd
Wolf-fish	*Anarhichas* sp.	4	bd	0.19
Edible sea urchin	*Echinus* sp.	4	0.09	0.34
Blue mussel	*Mytilus edulis*	14	0.09	0.69
Brown seaweed[a]	*Ascophyllum nodosum*	2	269	350[b]
Brown seaweed[a]	*Fucus vesiculosus*	1	-	233
Brown seaweed[a]	*Fucus serratus*	1	-	162
Snails	*Littorina* sp. and *Patella vulgata*	19	4.6	12.0

n=number of samples analysed
bd= below detection limit
[a]bulk sample; concentrations are given in Bq/kg dry weight
[b]average of 9 sub-samples

3.2 The distribution of ^{99}Tc in the brown seaweed *Ascophyllum nodosum*

A. nodosum is the only brown seaweed in Norway that is harvested commercially for the production of seaweed meal. Knowledge about the uptake and behaviour of ^{99}Tc in this species is therefore of special interest. Studies by Topcuoglu and Fowler[5] and Bonotto *et al.*[6] indicate that the uptake of ^{99}Tc in brown seaweeds takes place in young and metabolically active tissues. Moreover, it is shown that elevated temperatures and illumination enhance the accumulation rate (Topcuoglu and Fowler[5], Van der Ben *et al.*[7]).

This suggests that the uptake is an active and physiologically controlled process, and not only a passive absorption, although this remains to be shown.

Individual shoots of *A. nodosum* may reach 13 years of age, and from the age of 3, the main shoot and its branches develop one bladder per year (Baardseth[8]). It is therefore relatively simple to both determine the age of an individual shoot, and dissect it into yearly growth segments. In a preliminary study (Heldal *et al.*[9]), 9 replicate analyses of a bulk sample of *A. nodosum* gave a spread in the results from 275 to 393 Bq/kg d. w., with an average of 350 Bq/kg d. w. (Table 1). The spread of this data set is significantly higher than the analytical uncertainty of 5-10 %. We hypothesized that the spread was caused by an uneven distribution of ^{99}Tc in the bulk sample, and that the sub-samples actually contained different concentrations. Based on this, we decided to investigate the ^{99}Tc-concentration in yearly growth segments of *A. nodosum*. Preliminary results show that the ^{99}Tc-concentrations increase with the age of the growth segment (Table 2). One explanation for this may be that the concentrations in yearly growth segments reflect the seawater concentration in the year of tissue formation. Alternatively, the distribution pattern of ^{99}Tc in the annual shoot sections may reflect translocation of ^{99}Tc downwards in the plant, after uptake in the youngest sections, so that shoots act as sinks for 99Tc. This needs to be tested by analysing more samples.

The results are nevertheless of great importance for monitoring purposes. They may confirm that specific protocols for sample collection are needed when long-term studies are conducted and that care must be shown when comparisons are to be made between different studies.

Table 2. *^{99}Tc in yearly growth segments of the brown seaweed Ascophyllum nodosum*

	Bq/kg dry weight (d. w.)			
	Værlandet Apr-03	Rørvik Sep-03	Rørvik Sep-03	Espegrend Jan-06
2005-shoots				190
2004-shoots				263
2003-shoots	510	102	115	374
2002-shoots	688	188	198	451
2001-shoots	840	184	204	502
2000-shoots	-	218	247	502
1999-shoots	-	-	-	583
1998-shoots	-	-	-	644
1997-shoots	-	-	-	819
1996-shoots	-	-	-	964
receptacles	320	-	-	167

4 ACKNOWLEDGEMENTS

The Research Council of Norway (NFR) and the Institute of Marine Research (IMR) funded this study. Local fishermen, Directorate of Fisheries, Finnmark, and Værlandet School are greatly acknowledged for assistance with sample collection. HEH is grateful to Paul Blowers and Rachel Bonfield (CEFAS) for training in the analytical method and Penny L. Liebig and Ingrid Sværen for help with laboratory work.

References

1 B. R. Harvey, K. J. Williams, M. B. Lovett and R. D. Ibbett, 1992. *J. Radioanal. Nucl. Chem.*, **158**, 417.

2 J. E. Brown, A. K. Kolstad, A. L. Brungot, B. Lind, A. L. Rudjord, P. Strand and L. Føyn, 1999. *Marine Pollution Bulletin*, **38**, 560-571.

3 P. J. Kershaw, H. E. Heldal, K. A. Mork, A. L. Rudjord, 2004. Variability in the supply, distribution and transport of the transient tracer ^{99}Tc in the NE Atlantic. Journal of Marine Systems, **44**, 55-81.

4 P. J. Kershaw, D. McCubbin, K. S. Leonard, 1999. Continuing contamination of north Atlantic and Arctic waters by Sellafield radionuclides. The Science of the Total Environment, **237/238**: 119-132.

5 S. Topcuoĝlu and S. W. Fowler, 1984. Factors affecting the biokinetics of Technetium (^{95m}Tc) in marine macroalgae. Marine Environmental Research, **12**: 25-43.

6 S. Bonotto, V. Robbrecht, G. Nuyts, M. Cogneau and D. Van der Ben, 1988. Uptake of Technetium by Marine Algae – Autoradiographic Localization. Marine Pollution Bulletin, **19 (2):** 61-65.

7 D. Van der Ben, M. Cogneau, V. Robbrecht, G. Nuyts, A. Bossus and C. Hurtgen, 1990. Factors Influencing the Uptake of Technetium by the Brown Alga *Fucus serratus*. Marine Pollution Bulletin, **21 (2):** 84-86.

8 E. Baardseth, 1970. Synopsis of biological data on knobbed wrack *Ascophyllum nodosum* (Linnaeus) le Jolis. FAO Fisheries Synopsis No. 38 Rev. 1.

9 H. E. Heldal, K. Sjøtun, P. Roos, I. Sværen, P. L. Liebig, L. Føyn, 2005. The brown seaweed *Ascophyllum nodosum*: an archive for exposure of technetium-99 (^{99}Tc) over time? Pp.180-183 in: Proceedings from the 2nd International Conference on Radioactivity in the Environment, Nice, 2.-6. October 2005.

MEASURING THORON (^{220}Rn) IN NATURAL WATERS

W.C. Burnett[1], N. Dimova[1], H. Dulaiova[1], D. Lane-Smith[2], B. Parsa[3], and Z. Szabo[4]

[1]Environmental Radioactivity Measurement Facility, Department of Oceanography, Florida State University, Tallahassee, FL 32306, USA
[2]Durridge Co., Inc., 7 Railroad Avenue, Suite D, Bedford, MA 01730,USA
[3]New Jersey Department of Health and Senior Services, Radioanalytical Services, Trenton, New Jersey 08625
[4]U.S. Geological Survey, 810 Bear Tavern Road, W. Trenton, New Jersey 08628

1 INTRODUCTION

We have been using radon-in-air monitors coupled together with a water-air exchanger to measure ^{222}Rn in coastal waters as a tool to locate and quantify direct groundwater discharge into the sea. Recently, we began to investigate the possibility of making concurrent analysis of ^{220}Rn (thoron, $t_{1/2}$= 56 s) to "prospect" for points of groundwater entry. While the half-life of thoron makes its assay sensitive to variations in air and water flow rates, the short half-life is also an advantage because its detection ensures that one must be close to a source. Another useful application of thoron analyses in water is to locate radium-rich deposits that develop in oil/gas and water supply pipelines. This can be done via ^{220}Rn analysis at points along a pipeline, or at a single site while varying the water flow rate. Determination of the precise location of radioactive scale could avoid expensive and unnecessary remediation or at least affect distribution system maintenance programs. We present here the theoretical basis and some examples from public water supplies in New Jersey where some thorium-series radioactive decay products were already known to be elevated. We present a system for making continuous thoron-in-water measurements and show how one can estimate the volume, distance, and source strength of the contamination.

There are three naturally-occurring isotopes of radon: ^{222}Rn ("radon", $t_{1/2}$= 3.82 d, ^{238}U chain), ^{220}Rn ("thoron", $t_{1/2}$= 56 s, ^{232}Th chain), and ^{219}Rn ("actinon", $t_{1/2}$= 3.96 s, ^{235}U chain). Most attention concerning ^{222}Rn has centered on human health effects after inhalation and ingestion, however it also provides many useful geophysical and hydrologic tracer applications. For example, considerable efforts have been made over the years to use radon measurements in soil gas and groundwater for earthquake predictions. A particularly intriguing data set showed a peak in the groundwater ^{222}Rn 10 days before the disastrous 1995 earthquake in Kobe, Japan.[1] Radon has been used as a tracer of ground-water/surface-water mixing and matrix diffusion in fractured-rock aquifers.[2,3] We have been using ^{222}Rn measurements for coastal oceanographic applications. In particular, radon is an excellent tool for locating and quantifying the direct groundwater (high radon) discharge into coastal ocean waters (low radon). The principles and experimental approach for this application have already been described in a series of papers.[4,5,6] Gas exchange across the air-sea interface can also be studied experimentally using ^{222}Rn measurements.[7,8]

While most health investigations are concerned with radon activities in air, radon-in-water measurements may also have important human health implications. Degassing of the radon from the drinking-water supply to household air during ordinary domestic water use such as showering or washing laundry may be a substantial source of radon to indoor air prompting several states including New Jersey to evaluate "multi-media mitigation programs" for radon.[9,10,11] Recent studies at the University of Iowa have shown that build-up of radium-bearing scale or adsorption of radium onto corrosion products inside water distribution pipelines can act as a source for radon generation.[12,13,14] This can result in the unusual circumstance of having more radon appear at points along a water distribution pipeline than actually entered the system from the natural source.

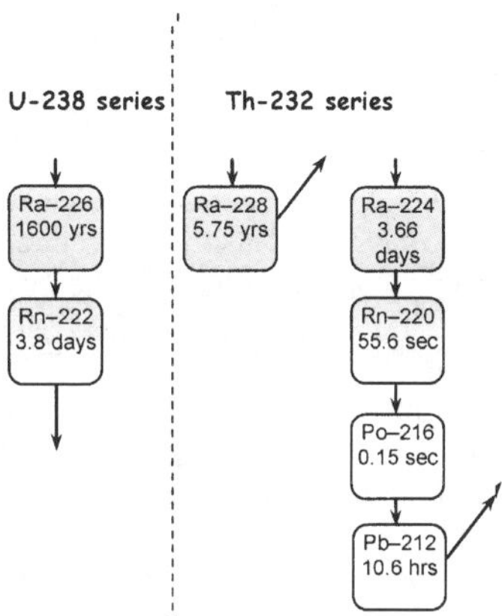

Figure 1 *Abbreviated ^{238}U and ^{232}Th decay chains illustrating the relative positions of radon (^{222}Rn) and thoron (^{220}Rn). The remaining natural isotope of radon, actinon (^{219}Rn), is a decay product of ^{223}Ra in the ^{235}U chain.*

Recently, we began considering the use of ^{220}Rn (Fig. 1) for investigations of environmental radioactivity in water supply systems as well as for possible coastal oceanographic applications. In both cases, the concept is that the very short half-life of thoron would ensure that its detection meant that one must be very close to a "source." In the case of groundwater flow into the coastal zone, the half-life of ^{222}Rn is sufficiently long that a radon anomaly could be carried hundreds of meters or even several kilometers away from a submarine spring by oceanic currents. Thoron, on the other hand, completely decays in about 5 minutes so a ^{220}Rn anomaly could only be present in coastal waters in the immediate vicinity of an active source. Such a source could be a submarine spring or a high concentration of thorium-bearing minerals (e.g., monazite) in seabed sediments. Studies of naturally-occurring radioactivity in groundwater in the United States have indicated that the occurrence of ^{228}Ra and ^{224}Ra is more widespread than previously known and in some states, such as New Jersey, water from principal aquifers used for drinking-water supply frequently have ^{228}Ra and ^{224}Ra concentrations greater than 5 pCi/L.[15,16] In a water or oil/gas pipeline, radon generated by radium deposits within the system would persist for days to weeks – longer than the residence time of waters in most pipeline systems. However, one could "prospect" for a ^{228}Ra/^{224}Ra source by analysis of thoron in water. Ideally, the analysis would be performed directly from points of entry to the pipeline and would produce immediate results. We present in this paper the measurement instrumentation, strategies employed in sampling, mathematical foundation for source model assumptions, and a pilot demonstration sampling project in a Ra–contaminated system in New Jersey.

2 EXPERIMENTAL

2.1 RAD-7 Instrumentation

Our basic approach for continuous analysis of radon and thoron in water is equilibration of these gases in a stream of flowing water with a stream of air which is being re-circulated through a commercial radon-in-air monitor (Fig. 2). We are using a RAD-7 (Durridge Co., 7 Railroad Ave., Suite D., Bedford, MA 01730) for the radon-in-air monitor because it is portable, durable, very sensitive, and operates in a continuous mode. The RAD-7 uses a high electric field with a silicon semiconductor detector at ground potential to attract positively charged radon daughters. The two polonium isotopes in the radon chain, $^{218}Po^+$ ($t_{1/2}$ = 3.10 m; alpha energy = 6.00 MeV) and $^{214}Po^+$ ($t_{1/2}$ = 164 μs; 7.67 MeV) provide a reading for the activity of ^{222}Rn while the thoron daughter, $^{216}Po^+$ ($t_{1/2}$ = 0.145 s; 6.78 MeV), provides an indicator of the decay of ^{220}Rn. Importantly, the RAD-7 has energy window settings that allow one to discriminate between all alpha-emitting polonium isotopes including the longer-lived ^{210}Po ($t_{1/2}$ = 138.4 d; 5.30 MeV) that invariably builds up over time due to accumulation of the long-lived beta-emitter ^{210}Pb ($t_{1/2}$ = 22.3 y). This discrimination feature creates a significant advantage in terms of sensitivity as the background for the specific alpha energy regions of interest will remain very low, close to zero. Furthermore, energy discrimination allows one to select either or both the ^{218}Po and ^{214}Po windows for ^{222}Rn assessment while monitoring the ^{216}Po window for thoron analysis. Radioactive equilibrium of the polonium daughters with ^{222}Rn will range from about 15 minutes for ^{218}Po to approximately 3 hours for ^{214}Po. The ^{214}Po lags behind because of the longer-lived intermediate beta-emitting daughters, ^{214}Pb ($t_{1/2}$ = 27 m) and ^{214}Bi ($t_{1/2}$ = 19.9 m). The thoron daughter, ^{216}Po, will reach radioactive equilibration almost instantaneously because of its very short half-life.

The water-air exchanger (commercially referred to as the "RAD-AQUA") is simply a clear plastic (acrylic) cylinder which has water flowing continuously through a narrow spray nozzle with a provision for a stream of air which is pumped, either from the built-in air pump in the RAD-7 or an external pump. The air is then circulated through a bed of desiccant and to the RAD-7 for measurement. A Nafion drying tube, commercially called a "Drystik," may be placed in the air stream to preserve the desiccant. After some time, the radon concentration in the air, which is being continuously recycled, reaches equilibrium with the radon in the water, the ratio at equilibrium being determined by the water temperature:

$$a' = 0.105+0.405e^{-0.0502T} \tag{1}$$

where a' is the concentration ratio of water to air (about 1:4 at room temperature), and T is the temperature of the water in degrees C.[17]

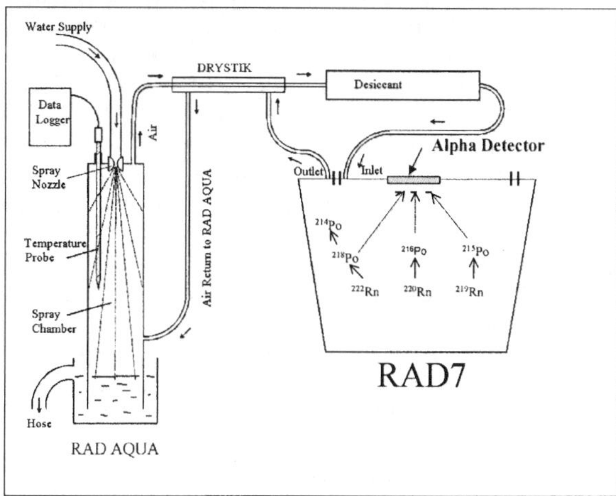

Figure 2 *Schematic of the "RAD-7/RAD-AQUA" system for continuous analysis of radon-in-water. The system as shown relies on the internal air pump of the RAD-7 and is thus more suitable for ^{222}Rn than ^{220}Rn analysis.*

Thus, the system is operated by providing a constant stream of water to be analyzed to the exchanger, continuously circulating the air through the closed loop (exchanger → desiccant → RAD7 → exchanger), obtaining radon/thoron-in-air measurements, and calculating the radon/thoron-in-water activity concentrations at equilibrium based on the temperature dependence and RAD-7 calibrations provided by the manufacturer. Once set up, this can all be done automatically. The water could be fed from a tap, a well, a submersible pump immersed in the ocean, or any other water source of interest. Depending upon the concentration and study requirements, the RAD-7 would be programmed to integrate counts (in either or both the ^{218}Po and ^{214}Po windows for ^{222}Rn; and ^{216}Po for ^{220}Rn) over whatever time period is necessary for the required precision, or desired for the application at hand. The data are stored in a data logger on-board the RAD-7 and easily downloaded to a PC for final analysis. There is a slight spillover of the lower-energy counts from ^{214}Po (called channel "C" in the RAD-7) into the thoron window (channel "B"; the ^{218}Po window is channel "A"). This spillover is generally less than 2%; (the unit we have been using, serial No. 1152, has a 1.4% spill) and thus does not pose a problematic bias except when the ^{214}Po (^{222}Rn) count rate is very much higher than that of ^{216}Po (^{220}Rn). This overlap is characterized during Durridge's calibration procedure and thus can easily be corrected for as all channels are continuously monitored.

2.2 Optimizing the System for Thoron Measurements

Because of its short half-life, the assay of thoron is very sensitive to variations in air and water flow rates. The response time of the system depends on the half lives of the radon isotopes and their respective polonium daughters, the volume of the air loop, the speed of transfer of radon and thoron from the water to the air (which depends on the efficiency of the aeration, likely due to a number of parameters such as droplet size, etc.), the flow rate of the re-circulating air, the volume of water in the exchanger, and the flow rate of water to the exchanger.[18] The half-life of ^{216}Po, 0.145 s, dictates an ultimate theoretical limit, for the 95% response time, of only about 0.7 seconds, assuming everything else was instantaneous. Since there is about four times more radon/thoron in the air phase than the aqueous phase at equilibrium, at least four times more water than air must flow through the system to deliver all the radon that is required to maintain equilibrium. Again, that is assuming everything is working at maximum efficiency which is unlikely.

Of all the possible parameters which may effect the equilibration time, the only ones which could easily be controlled to some extent are the flow rates of the re-circulated air and water. In our previous experimental work with ^{222}Rn, we had relied on the internal air pump of the RAD-7 which is fixed at a flow rate of about 1 L/min. Since the very short half-life of thoron dictates that we perform the analysis as quickly as possible, we have now installed an external air pump so we can vary the flow rate of the air as well as the water (Fig. 3). We report here our observations on the effect of different water and air flow rates on the response to a constant source of thoron.

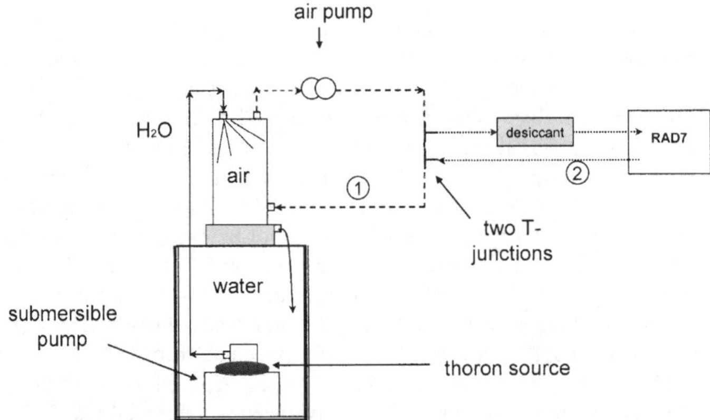

Figure 3 *Experimental set-up for optimizing the counting system for thoron measurements. The "thoron source" is made from acrylic fiber impregnated with MnO_2 which is used as a thorium/radium adsorber. There are two closed air loops: (1) circulating through the exchanger driven by an external air pump; and (2) a separate loop between the RAD-7 and the primary air path driven by the internal pump of the RAD-7.*

We produced a thoron source for laboratory experiments by adsorbing thorium (using a solution made from a ^{232}Th sulfate compound known to be over 40 years old so all daughters would be in secular equilibrium) onto a MnO_2 impregnated acrylic fiber. Such fiber, called "Mn fiber" by oceanographers, has very high adsorptive capacities for radium and thorium and is thought to have a high emanation factor for radon.[19] This fiber was submerged in a ~50-liter deionized and radium-free water reservoir and water was pumped from a variable-speed submersible pump to the modified exchanger/RAD-7 system.

Using water flow rates ranging from about 1 to 5 L/min (the highest the system would produce) and three different air flow rates (0.8, 3.0, and 7.0 L/min), we can see the clear dependence of the RAD-7 count rate in the thoron channel resulting from both these parameters (Fig. 4). The sensitivity of the response (cpm/water flow rate) more than doubles by increasing the air flow from the lowest setting to the highest. Note also that the higher the air flow, the lower the water flow can be before a response is detected, i.e., the y-intercept is highest at the lowest air flow. The slope is greatest at the highest airflow, again indicating increased sensitivity.

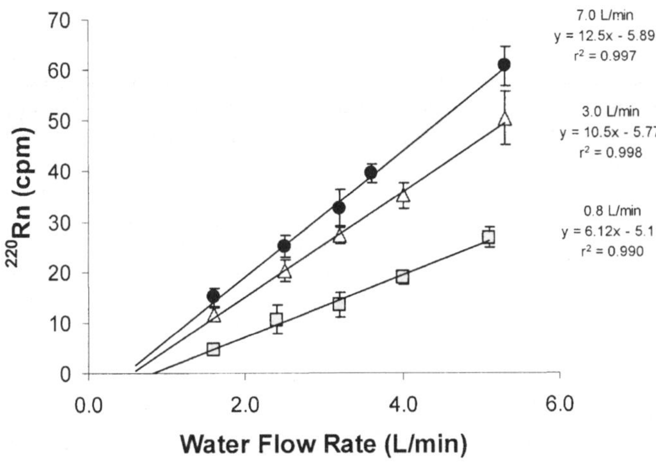

Figure 4 *Response in the thoron channel from a constant thoron source as functions of water and air flow rates. Water flow rates are shown on the x-axis and the air flow rates shown are for 0.8 (boxes), 3.0 (triangles), and 7.0 (circles) L/min.*

3 THORON AS A PROSPECTING TOOL

3.1 Background and General Strategy

We suggest that thoron analyses in water may be used to locate radium-rich deposits that develop in oil/gas and domestic water distribution pipelines. This source identification could be done via ^{220}Rn analysis at different points along a pipeline, or at a single site while varying the total water flow rate in the pipeline. Knowledge of the precise location of radioactive scale could help avoid expensive and unnecessary remediation. We measured radon and radium isotopes in some public water supplies in New Jersey where thorium-chain radioactive decay products were already known to be elevated. In particular, several older water supply systems in west central and southern New Jersey have been shown to have elevated gross alpha anomalies and high unsupported ^{212}Pb activities.[16,20,21,22] The observance of unsupported ^{212}Pb ($t_{1/2}$=10.64 hr) was attributed to the presence of ^{220}Rn. Considering the short half life of ^{220}Rn, within a few minutes, all its initial concentration will decay to ^{212}Pb. Consequently, the detection of unsupported ^{212}Pb could serve as an indicator for the presence of ^{220}Rn. However, due to its comparatively longer half life, ^{212}Pb will be transported further along the distribution system and detected beyond the proximity of its source. In many of these contaminated water distribution systems, radium isotopes were present but not significantly enriched in the water. We hypothesized that radium, while not particularly high in these waters, could be concentrated in pipelines either by scale formation or more likely by adsorption onto corrosion products such as hydrous Fe/Mn oxides. While this process may take on the order of years to decades to develop a significant concentration, the adsorbed radium would eventually become a radon-thoron generator inside the pipeline. Whether the adsorber generated more radon or thoron would depend upon the $^{226}Ra/^{228}Ra$ ratio in the source water (and thus in the deposit). Adsorption of radium would be especially effective if the corrosion product consisted of oxides of iron or manganese. MnO_2, in particular, is known to be an excellent concentrator of radium with a distribution coefficient, K_D of greater than 10^4.[23,24] Thus, we feel that in water supply systems, hydrous Fe/Mn oxides are most likely to exhibit significant sequestration of radium and release of gaseous decay products. Other types of mineral scale ($CaCO_3$, $CaSO_4$, etc.) that are commonly found in drinking water systems would normally contain too little radium (very low Ra/Ca ratios in typical aquifers) and the crystal structures would inhibit radon emanation.[25] We thus suggest that the radium-rich deposits that develop in water supply pipelines are fundamentally different than the high-activity $BaSO_4$ scale known to occur in some oil/gas pipelines.[26] In either case, the thoron prospecting approach could be used to locate such deposits.

Should naturally-occurring radium become enriched within solid deposits inside pipelines, there is a high potential that this would eventually produce a source of radon isotopes and associated decay products to the water. As gases radon and thoron may easily migrate out of the solid products and into the water. While ^{222}Rn may be considered more important in terms of health effects, ^{220}Rn would be the best indicator that one is close to a such "radioactive source" since its short half-life would cause it to decay very quickly (~97% would decay within 5 minutes). Thus, one must be close to its source in order to see it before it decays. Progeny of the ^{220}Rn decay, such as ^{212}Pb, might be detectable as elevated gross alpha-particle or gross beta-particle activity, causing further concern about the location of the radioactivity deposit.

Our goal in making these thoron measurements is to be able to estimate the distance and source strength of the contamination. In this manner we could identify those sections of pipeline where contamination exists and avoid the cost of needless remediation of uncontaminated portions. Alternatively, the information could be used in developing distribution system maintenance and corrosion control programs. We present below the theoretical basis for interpreting thoron-in-water measurements in two situations: (1) where the water flow through our detection system is insignificant compared to the total water flow in the pipeline; and (2) where the only water flow is that through the detection system. In the former case, the ^{220}Rn activity in the water will be constant as we have not disturbed the interaction between the solid radium source and the flowing water (the "infinite source" model). In the latter situation, the changing water flow rate within our exchanger would effect the activity of thoron in the water as it represents the only flow (the "finite source" model).

3.2 Infinite Source Model

In this case, we assume that the water flow through our detection system (a few liters per minute) is negligible compared to the much higher flow in the pipeline. This situation could occur under a few different circumstances. For example, we could tap into a main pipeline that has a very high flow such that the small volumes we draw are insignificant and would not influence the concentration of thoron emanating off a source within that pipeline. Another possibility would be that the water system distribution point we are sampling from is uncontaminated but is an offshoot of a contaminated water main. This would be the case, for example, when an old water pipeline in the street serves a secondary, much smaller pipeline that goes to a private residence. The main pipeline may have thoron contamination that could be brought into the house when the tap is turned on. If the water flow in the main is constant, thoron contamination (emanating from a nearby source) in the water should also be steady-state (decay of ^{220}Rn balances its production) as it passes the tap that serves the house.

This approach can be illustrated by an example we encountered from an outside tap at a private house in Bridgeport, New Jersey. The New Jersey Department of Health and Senior Services (NJDHSS) had already seen elevated levels of unsupported ^{212}Pb (a decay product in the thoron chain) and high, short-lived gross alpha activities within this water supply system. The elevated gross-alpha activity exhibited a decay rate of about 11 hours, and was attributed to the alpha-emitting ^{212}Pb decay products ^{212}Bi and ^{212}Po. The ^{224}Ra and ^{226}Ra activities were determined to be negligible. We measured sufficiently high levels of ^{220}Rn at this site to experiment with different flow rates through the exchanger in order to test possible relationships between water flow rate and thoron activity. We did, in fact, observe a very high correlation between the activity of thoron at the point of measurement and the water flow rate when we varied the water flow through our detection system (Fig. 5a). We interpret this to mean that we were at some fixed distance from the source of ^{220}Rn (probably the water main in the street) so we received higher activities at higher flow rates because of less decay during transit. Having made these observations, it should be possible to interpret the results in terms of estimating both the magnitude (thoron activity in the water at the source) and distance to that source.

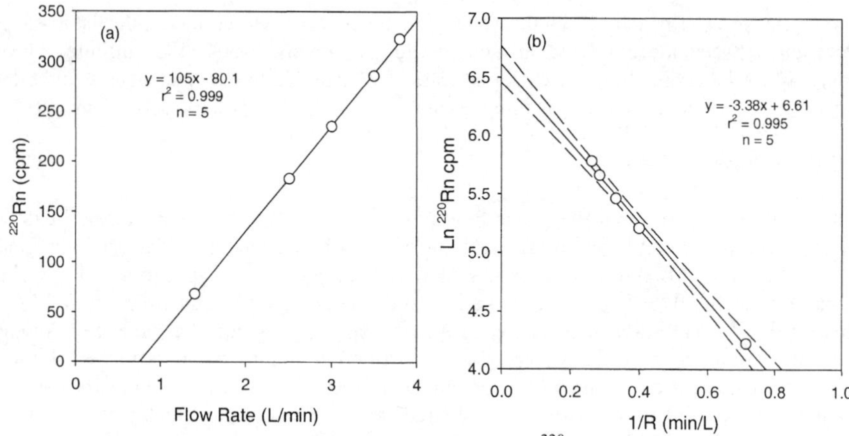

Figure 5 *(a) Variation of the measured concentration of* ^{220}Rn *at an outside faucet at a private house in Bridgeport, New Jersey, as a function of the water flow rate. (b) The natural logarithm of the* ^{220}Rn *count rate versus the inverse of the water flow rate can be used to assess the activity of the thoron in water at the source and its volume (distance) away from the measurement. The dashed lines represent the 95% confidence intervals of the regression.*

For the analysis, we assume that the diameter of the pipe serving the house is constant (¾-inch inside diameter); that the tap into the water main in the street represents a "point" source, and that the only reason for a variation in ^{220}Rn activity with flow rate is because of differences in decay (travel time from source to point of measurement). We define the following terms:

$^{220}Rn^{o}$ = steady-state activity of unsupported ^{220}Rn in the water at the source, Becquerels per liter (Bq/L)
^{220}Rn = activity of unsupported ^{220}Rn in the water at the detection system (Bq/L)
V = volume of water in the pipeline between the source and our detector, liters (L)
R = flow rate of water through the exchanger, liters per minute (L/min)
t = transit time for the water to flow from the main to the detection system (min)
λ = decay constant of ^{220}Rn (λ = Ln $2/t_{½}$; 0.756 min^{-1} or 0.0126 s^{-1})

When flow occurs, the initial ^{220}Rn activity concentration in the water coming from the point source would decay exponentially as it advances towards our detector:

$$^{220}Rn = {}^{220}Rn^{o} \bullet e^{-\lambda t} \qquad (2)$$

Since the transit time, t, to reach the detector may be expressed as V/R, we can write:

$$^{220}Rn = {}^{220}Rn^{o} \bullet e^{-\lambda V/R} \qquad (3)$$

We can then take the natural logarithm of each side of the equation:

$$Ln^{220}Rn = Ln^{220}Rn^{o} - \lambda V \frac{1}{R} \tag{4}$$

This is in the form of an equation of a straight line (y = mx + b; where m = slope; and b = y-intercept) in the space of the logarithm of the ^{220}Rn activity and 1/R (Fig. 5b). Using this approach, the slope of the best-fit linear regression line would equal $-\lambda$V, allowing us to calculate V. Since we know the decay constant and have an analysis of the slope from our observations (slope = -3.35 L/min), we can now calculate the volume in between the source and our detection system:

$$\mathrm{V} = \frac{\mathrm{slope}}{\lambda} \tag{5}$$

In this example, the volume equals (3.35 L/min/0.756 min^{-1}) which equals 4.43 liters. Since we are assuming that the inside pipe diameter is constant at 0.75 inches (this was the diameter where we tapped into the system), we can calculate ($V = \pi r^2 l$, where l = length) that each meter of pipe is equivalent to 0.285 liters. Thus, the theoretical point source must be located 4.43 L/0.285 L/m or 15.5 m away from our detection system. One may also note from Figure 5a that at flow rates below 0.76 L/min, there is no thoron detected by the system. If we assume that this represents about 6 half-lives (5.6 min or >98% decay) of ^{220}Rn decay, then V = 5.6 min x 0.76 L/min or 4.3 L, essentially the same result as the prior analysis.

The regression line in Figure 5b also provides an indication of the initial activity of thoron in the water. The y-intercept (6.61 on a Ln scale) represents the count rate when the inverse of the flow rate is at zero (the flow rate is at infinity), thus the "instantaneous" count rate would be exp(6.61) or 742 cpm. Correcting for the solubility of radon in water (a' = 0.316) at the in situ water temperature (13°C) and using a rated RAD-7 efficiency for thoron of 5.68 cpm/(Bq/L) [0.210 cpm/(pCi/L)], this translates to a thoron activity of 41.3 Bq/L in the water. This should represent the activity of ^{220}Rn about 15 m upstream in the pipeline from the point of measurement.

3.3 Finite Source Model

In this case, the water being measured is in the same pipeline as the contamination and the only flow in that pipeline is that through our detection system. Thus, the water flow rate through the exchanger would influence the activity of thoron in the water. This assumes that the source is a radium-bearing solid that has a constant emanation rate of thoron into the water. So increasing the water flow rate would decrease the activity concentration in the water. We define one additional term, S, as the effective source strength of the ^{224}Ra contamination (Bq). All other definitions are as given previously. The thoron activity concentration in the water at the source would be:

$$^{220}Rn^{o} = \frac{\lambda S}{R} \tag{6}$$

Since it takes V/R minutes to reach our detection system from the source, the ^{220}Rn will decay during transit to:

$$^{220}Rn = \frac{\lambda S}{R} \bullet e^{-\lambda(V/R)} \tag{7}$$

We can then take the natural logarithm of the equation:

$$Ln^{220}Rn = Ln\left(\frac{\lambda}{R} \bullet S\right) - \frac{\lambda}{R} \bullet V \tag{8}$$

Which is equivalent to:

$$Ln^{220}Rn = Ln\left(\frac{\lambda}{R}\right) + Ln(S) - \frac{\lambda}{R} \bullet V \tag{9}$$

And can also be written as:

$$Ln^{220}Rn - Ln\frac{\lambda}{R} = -V \bullet \frac{\lambda}{R} + Ln(S) \tag{10}$$

In this case, we may plot [Ln(^{220}Rn)-Ln(λ/R)] versus λ/R and a straight line with a slope of –V and an intercept of Ln(S) would result. Thus, one may calculate the volume (distance) from the source as well as the source strength.

A series of thoron measurements at different water flow rates collected at a private residence in Florence, New Jersey appears to fit this model. Again, this water supply system had shown evidence of elevated gross alpha-particle activity and unsupported ^{212}Pb, indicating a possible ^{228}Ra-thoron source somewhere inside the pipeline. In this case, the plot of the RAD-7 response (cpm) versus water flow rate shows curvature at higher flows suggesting that the contamination was in the same pipeline and that this finite source model may be more appropriate (Fig. 6a). We transposed the results as shown above, converted each data point into activity units using the temperature-dependent solubility coefficient and the calibration of the instrument. The resulting plot is very linear (Fig. 6b). The slope indicates a volume from the source of about 10 liters. The y-intercept indicates a source strength of exp(3.87) or 48 Bq. Note that one could also plot the [Ln(^{220}Rn)-Ln(λ/R)] function in terms of the RAD-7 response (cpm) instead of calculating activity for each data point and do the conversion from cpm to activity at the end. The result would be the same.

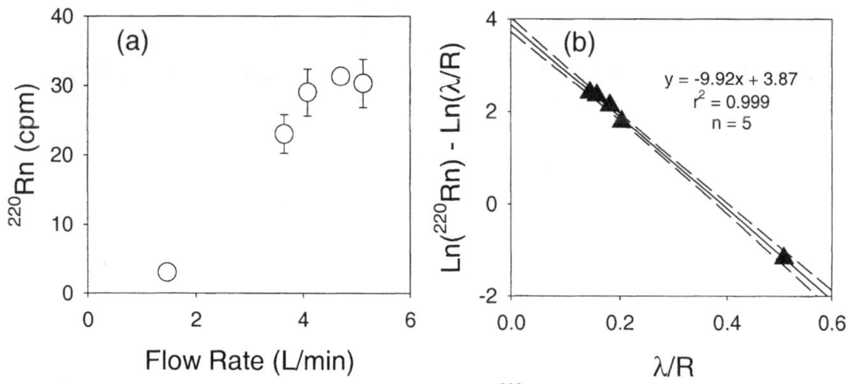

Figure 6 *(a) Variation of the measured count rate of ^{220}Rn at an outside faucet at a private house in Florence, New Jersey, as a function of the water flow rate.*
(b) Transposing the data in the manner described allows one to assess the activity of the thoron source (y-intercept = LnS) and the volume of water between the source and detection system (slope = -V). Units for the x- and y-axes are volume (liters) and activity (Bq), respectively. The dashed lines represent the 95% confidence intervals of the regression.

4. SUMMARY/FUTURE WORK

We have shown that ^{220}Rn (thoron) can be detected in water using a RAD-7/RAD-AQUA system. The efficiency of the detection is enhanced by use of an external air pump. By establishing a relationship between the water flow rate through the exchanger and the RAD-7 response, one can estimate the volume/distance to the source and extrapolate back to an "instantaneous" thoron activity as seen by the instrument. Using such a system, one can prospect for high radium accumulations within pipelines. This would be useful for assessing the presence of radium-rich scale in oil and gas pipelines and radium adsorbed to corrosion products in domestic water supply systems. The measurements are simple, mostly automated, and inexpensive. Defining more precisely the locations of radium contaminations could avoid unnecessary remediation costs. Such information could also help target remediation to where it is most needed, or could be useful in designing distribution system maintenance and corrosion control practices. The research also indicates that the occurrence of unsupported ^{212}Pb in the distribution system may also be controlled more by the location of ^{220}Rn sources within the system than within the aquifer.

We hope to develop the sensitivity of the measurements even further in order to provide a tool for coastal oceanographic applications such as searching for submarine springs. With concentrations likely well below 0.1 Bq/L, this remains a challenging but potentially valuable addition.

Acknowledgements

The authors thank Stephen Jenniss from the New Jersey Department of Health and Senior Services and Barker Hamill and Lee Lippincott from the New Jersey Department of Environmental Protection (NJDEP) who provided support and made arrangements for our sampling programs. Pamela Reilly and Nicholas Smith from the USGS provided useful water quality information and assisted in the sampling. We also thank the personnel at the Florence, Pennsgrove, and Bridgetown, New Jersey water supply facilities for allowing us to make these measurements. Vincent dePaul provided constructive comments concerning this research. Partial financial assistance for this research was provided by the PG Research Foundation (PGRF), by NJDEP, and the USGS. Any use of trade, product, or firm names in this paper is for descriptive purposes only and does not imply endorsement by the U.S. Government or the State of New Jersey.

References

1 G. Igarashi, S. Saeki, N. Takahata, K. Sumikawa, S. Tasaka, Y. Sasaki, M. Takahashi, and Y. Sano, *Science*, 1995, **269**, 60.
2 C.Bertin, and A.C.M. Bourg, *Environmental Science Technology*, 1994, **28**, 794-798.
3 W.W. Wood, T.F. Kraemer, and A. Shapiro, *Ground Water*, 2004, **42**, 552-567.
4 W.C. Burnett, G. Kim, and D. Lane-Smith, *Jour. Radioanal. Nucl. Chem.*, 2001, **249**, 167.
5 W.C. Burnett, and H. Dulaiova, *Jour. Environ. Radioact.*, 2003, **69**, 21.
6 H. Dulaiova, R. Peterson, W.C. Burnett, and D. Lane-Smith, *Jour. Radioanal. Nucl. Chem.*, 2005, **263(2)**, 361.
7 W.S. Broecker, J. Cromwell, and Y.-H. Li, *Earth Planet. Sci. Lett.*, 1968, **5**, 101.
8 H. Dulaiova, and W.C. Burnett, *Geophys. Res. Lett.*, 2006, **33**, L05606, doi:10.1029/2005GL025023.
9 C.T. Hess, M.A. Vietti, and D.T. Mage, *Environmental Geochemistry Health*, 1987, **9**, 68-73.
10 P.F. Folger, P. Nyberg, R.B. Wanty, and E. Poeter, *Health Physics*, 1994, **67**, 244-252.
11 V.T. dePaul, P.L. Gardner, A.J. Kopera, H.C. Roy, and Z. Szabo, Proceedings of the National Ground Water Association, Minneapolis, Minnesotta, June 7-8, 2000.
12 R.W. Field, E.L. Fisher, R.L. Valentine, and B.C. Kross, *Am. Jour. Public Health*, 1995, **85(4)**, 567.
13 E.L. Fisher, L.J. Fuortes, J. Ledolter, D.J. Steck, and R.W. Field, *Health Physics*, 1998, **74(2)**, 242.
14 E.L. Fisher, L.J. Fuortes, R.L. Valentine, M. Mehrhoff, and R.W. Field, *Environ. Internat.,* 2000, **26**, 69.
15 M.J. Focazio, Z. Szabo, T.F. Kraemer, A.H. Mullin, T.H. Barringer, and V.T. dePaul, *U.S. Geological Survey Water Resources Investigations Report,* 2001, **00-4273**.
16 Z.Szabo, V.T.dePaul, T.F. Kraemer, and B.Parsa, *U.S. Geological Survey Scientific Investigations Report* **2004-5224**.
17 F. Weigel, Chemiker Zeitung, 1978, **102**, 287.
18 D. Lane-Smith, S. Shefsky, Proceedings of the American Association of Radon Scientists and Technologists, Las Vegas, Nevada, November 7-10, 1999.

19 W.S. Moore, *Deep-Sea Res.*, 1976, **23**, 647.
20 B. Parsa, *Jour. Radioact. Radiochem.*, 1998, **9**, 41.
21 B. Parsa, W. Nemeth and R. N. Obed, *Jour. Radioact. Radiochem.*, 1999, **11**, 11.
22 B. Parsa, R.N. Obed, W.K. Nemeth, and G. Suozzo, *Health Physics*, 2004, **86**, 145.
23 D.S. Moon, W.C. Burnett, S. Nour, E.P. Horwitz, and A. Bond, *Applied Radiat. Isotopes*, 2003, **59**, 255.
24 S. Nour, A. El-Sharkawy, W.C. Burnett, and E.P. Horwitz, *Applied Radiat. Isotopes*, 2004, **61**, 1173.
25 G.J. White, A.S. Rood, *Jour. Environ. Radioact.*, 2001, **54**, 401.
26 I.S. Hamilton, M.G. Arno, J.C. Rock, R.O. Berry, J.W. Poston, J.R. Cezeaux, and J.M. Park, *Health Physics*, 2004, **87**(4), 382.

THE DETERMINATION OF GROSS ALPHA AND GROSS BETA ACTIVITY IN SOLIDS, FILTERS AND WATER - VALIDATION OF DUTCH PRE-NORMS

P.J.M. Kwakman[1], E. van der Graaf[2], P. de Jong[3]

[1] RIVM, PO Box 1, 3720 BA, Bilthoven, Netherlands
[2] KVI, Zernikelaan 25, 9747 AA, Groningen, Netherlands
[3] NRG-Arnhem, PO Box 9034, 6800 ES, Arnhem, Netherlands

1 INTRODUCTION

Four Dutch norms, important for the determination of alpha or beta emitters, have been validated. Gross-alpha and gross-bèta measurements are very useful for fast and easy screening of environmental samples after a nuclear accident.
The validation concerns the following norms:

- NVN 5622 – gross alpha in solids (thick source method) [1]
- NVN 5627 – gross beta in solids [2]
- NVN5636 – gross alpha and gross beta on filters [3]
- NEN 6421 – gross beta in water [4]

Commissioned by the Netherlands Normalisation-institute, RIVM has carried out this validation together with the Kernfysisch Versneller Instituut (Groningen, NL) and the Nuclear Research and Consultancy Group (Arnhem, NL).
The pre-norm texts were improved and modernized. The analytical characteristics have been determined in an intercomparison run carried out by six to ten laboratories. These characteristics include repeatability, accuracy, reproducibility and detection limit.

2 METHODS - NVN 5622 AND NVN 5627

2.1. General

All samples were counted with a Berthold LB 770 gas-flow counter in the alpha/beta mode, meaning both alpha and beta activities were measured in the same run.

2.2. NVN 5622 – gross alpha in solids

$CaSO_4$ powders were spiked with low and high amounts of ^{241}Am. These powders were checked for homogeneity with gamma spectrometry and sent to the participating laboratories. ^{241}Am was used both for calibration and spiking.

2.3. NVN 5627 – gross beta in solids

Mixtures of KCl and $CaSO_4$ powders were used as such or spiked with ^{90}Sr. These powders were checked for homogeneity with LSC and sent to the participating laboratories. ^{40}K was used for calibration.

3 RESULTS - NVN 5622 AND NVN 5627

3.1. NVN 5622 – gross alpha in solids

In the NVN 5622 intercomparison, seven laboratories analyzed three samples in triplicate (see table 1). The results of two labs were eliminated as outliers. Statistical analyses were performed using ISO 5725 [5] and NEN 7777 [6].

Table 1 *Results of NVN 5622 intercomparison*

Characteristic	Sample[a]		
NVN 5622 – gross alpha (Bq/g)	1	2	3
Average result ($\bar{a}_m$)	2,56	49,39	48,28
True value	2,76	53,99	55,74
Repeatability standard deviation (s_r)	0,10	2,01	2,19
Reproducibility standard deviation (s_R)	0,34	5,00	5,96
Systematic error (δ)	-0,20	-4,61	-7,46
Detection limit (AG_r)	0,20		

[a] sample 1: $CaSO_4$-powder with a low alpha-activity ((2,76 ± 0,06) Bq/g, ^{241}Am)
sample 2: $CaSO_4$-powder with a high alpha-activity ((54,0 ± 1,2) Bq/g, ^{241}Am)
sample 3: $CaSO_4$-powder with a high alpha-activity ((55,7 ± 1,2) Bq/g, ^{241}Am) and bèta-acivity (see NVN 5627)

3.2. NVN 5627 - gross beta in solids

In the NVN 5627 intercomparison, eight laboratories analyzed three samples in triplicate (see table 2). Four results were eliminated as outliers.

Table 2 *Results of NVN 5627 intercomparison*

Characteristic	Sample[a]		
NVN 5627 – gross beta (Bq/g)	1	2	3
Average result ($\bar{a}_m$)	10,48	37,48	5,26
True value	11,97	43,65	4,39
Repeatability standard deviation (s_r)	0,21	0,75	0,12
Reproducibility standard deviation (s_R)	0,30	1,13	0,82
Systematic error (δ)	-1,49	-6,17	0,87
Detection limit (AG_r)	0,046		

[a] sample 1: $CaSO_4$-powder with low bèta-activity ((12,0 ± 0,3) Bq/g, KCl and ^{90}Sr)
sample 2: $CaSO_4$-powder high bèta-activity ((43,7 ± 1,0) Bq/g, KCl and ^{90}Sr)
sample 3: $CaSO_4$-powder with a low bèta-activity ((4,4 ± 0,1) Bq/g, ^{90}Sr) and alpha-activity (see NVN 5622)

3.3. Discussion NVN 5622 (gross-alpha) and NVN 5627 (gross-beta)

In the gross-alpha intercomparison a relative reproducibility of 10-12 % was found. This is probably due to the preparation of a ^{241}Am-spiked $CaSO_4$-powder for calibration. For some laboratories who were not allowed to have an ^{241}Am solution this presented practical problems.
In the gross-beta intercomparison a trueness of -12 to -14 % was found in samples 1 and 2. This is due to the fact that the calibration is carried out with ^{40}K and the samples are spiked with both ^{40}K and ^{90}Sr. The average beta-energy of ^{40}K is higher than the average beta-energy of ^{90}Sr + ^{90}Y . Calculating a gross-beta activity with a too high efficiency leads to results which are too low. On the contrary in sample 3, a sample with a high alpha activity compared to beta activity, alpha-to-beta spill-over leads to a much *higher* beta activity. This is of course strongly dependent on the gross-alpha activity in the sample.

4 METHOD - NVN 5636

Glass fibre filters and membrane filters (Ø 50 mm) were spiked with weighed amounts of an ^{241}Am or ^{137}Cs standard solution. The spiking procedure was carried out by immersing the filter in a small volume of the standard solution and evaporation of the solvent. Homogeneity was checked with gamma spectrometry. ^{241}Am and ^{90}Sr were used for calibration.
The gross-alpha and gross-beta activities were measured separately with a gas flow counter: there were no alpha-to-beta spill-over issues.

5 RESULTS - NVN 5636

5.1. Gross alpha on filters

In the NVN 5636 intercomparison, six laboratories analyzed glass-fibre filters and membrane-filters for gross alpha and gross beta activity (see table 3 and table 4). The result for one laboratory was rejected as an outlier.

Table 3 *Results of NVN 5636 intercomparison (gross alpha)*

Characteristic Bq	Glassfibre[a]	Membrane filter[a]
Average result ($\overline{A}_m$)	22,9	22,8
True value	31,02	30,9
Repeatability standard deviation (s_r)	0,93	1,14
Reproducibility standard deviation (s_R)	0,67	1,28
Systematic error (δ)	-8,0	-8,2
Detection limit (AG_r)	0,022	0,014

[a] glassfibre-filter spiked with ^{241}Am (31,02 ± 0,09 Bq) and ^{137}Cs (46,3 ± 0,11 Bq)
membrane-filter spiked with ^{241}Am (30,9 ± 0,2 Bq) and ^{137}Cs (45,6 ± 0,4 Bq)

5.2. Gross beta on filters

Table 4 *Results of NVN 5636 intercomparison (gross beta)*

Characteristic Bq	Glassfibre[a]	Membrane filter[a]
Average result ($\overline{A}_m$)	35,2	42,4
True value	46,3	45,6
Repeatability standard deviation (s_r)	0,63	1,31
Reproducibility standard deviation (s_R)	4,6	5,9
Systematic error (δ)	-11,1	-3,3
Detection limit (AG_r)	0,074	0,081

[a] See note at bottom of table 3.

5.3. Discussion NVN 5636

Gross-alpha

Preparing spiked filters presented some problems. The spike solution was distributed throughout the filter. This resulted in an almost equal countrate with the filter in the 'normal' position or turned upside down. As the efficiency was determined with a one-sided spiked filter most of the reported activities were too low. In general, the determination of the correct efficiency for counting gross–alpha on filters is a problem. Both selfabsorption of alpha's in the filter and sampling of airdust will decrease the alpha-efficiency in an unknown but significant way. In this case, the 'bias' is exactly known and a correction factor can be calculated.

Gross-beta

The negative bias was caused by using ^{90}Sr/ ^{90}Y for the determination of the beta-efficiency and ^{137}Cs for spiking the filters. The average beta-energy of ^{90}Sr/ ^{90}Y is higher than the energy of ^{137}Cs and thus a too low activity is calculated.

6 METHOD - NEN 6421

Water sampled from river Waal was used for gross-beta measurements. Sample 1 is 'blank' Waal-water and to sample 2 and 3 a known amount of ^{137}Cs was added. Homogeneity was checked with gamma spectrometry. ^{40}K was used for calibration.
All samples were evaporated to dryness on a stainless steel planchette and counted with a gas flow counter.

7 RESULTS – NEN 6421

7.1. NEN 6421 gross-beta in water

In the NEN 6421 intercomparison ten laboratories analyzed three samples (see table 5). The results of one laboratory were rejected as an outlier.

Table 5 *Results of NEN 6421 intercomparison (gross beta in water)*

Characteristic	Sample[a]		
Bq/l	1	2	3
Average result ($\bar{c}_A$)	0,58	6,6	13,8
Reproducibility standard deviation (s_R)	0,17	1,0	1,4
Detection limit (AG_r); counting time 3600 s	0,035		
After correction for contribution of sample 1[b]			
Average result ($\bar{c}_A$)		5,9	13,3
True value		10,9	18,5
Recovery, in %		55	72

[a] sample 1 : Water from river Waal
sample 2 : As sample 1 with addition of (10,89 ± 0,05) Bq/l bèta-activity (^{137}Cs)
sample 3 : As sample 1 with addition of (10,00 ± 0,05) Bq/l bèta-activity (^{137}Cs) and (8,50 ± 0,02) Bq/l bèta-activity (^{40}K)
[b] : sample 2 and 3 (corrected): results of sample 2 and 3 after correction for bèta-activity in sample 1

7.2. Discussion NEN 6421

As the river Waal-water contains circa 0,58 Bq/l beta-activity the results of sample 2 and 3 were corrected for this 'background' activity. These corrected results are given in the bottom part of Table 5.
Similarly to the beta results on the filters is the negative bias caused by using ^{40}K for calibration and ^{137}Cs (sample 2) or both ^{137}Cs and ^{40}K (sample 3) for spiking the sample.

8 CONCLUSIONS

This paper describes the analytical characteristics of four Dutch norms, dealing with the determination of gross alpha and gross beta activity in solids, filters and water [7]. A gross alpha activity is reported as an '^{241}Am-equivalent' and a gross beta activity as a '^{40}K-equivalent' (of course in the case where ^{241}Am and ^{40}K are calibration nuclides).
The alpha or beta energy of the radionuclide in the sample may differ largely from the energy of the calibration nuclide. This will result in an unknown under- or overestimation of the gross alpha or gross beta activity. Therefore, these methods are not very accurate. On the other hand, these methods are mainly meant for screening purposes and processing large amounts of samples. In those cases, simplicity and fast sample preparation are more important than a high accuracy.

References

1 NEN 5622: 2006 - Radioactivity measurements - Determination of massic gross-alpha activity of a solid counting sample by the thick source method. NEN, Delft, The Netherlands (in Dutch).
2 NEN 5627: 2006 - Radioactivity measurements - Determination of massic gross-beta activity and massic rest-beta activity of a solid counting sample. NEN, Delft, The Netherlands (in Dutch).
3 NEN 5636: 2006 - Radioactivity measurements - Determination of artificial gross-alpha activity, artificial gross-beta activity and gamma spectometry of air filters and the calculation of the volumic activity of the sampled air. NEN, Delft, The Netherlands (in Dutch).
4 NEN 6421: 2006 - Water – Determination of volumic gross-beta activity and volumic rest-beta activity of non-volatile compounds. NEN, Delft, The Netherlands (in Dutch).
5 ISO 5725 : 1994, Accuracy (trueness and precision) of measurement methods and results – Parts 2, 3, 4, and 6.
6 NEN 7777 (2003), Environment – characteristics of analytical methods. NEN, Delft (in Dutch).
7 PJM Kwakman, ER van der Graaf and P de Jong. The validation of the Dutch pre-norms NVN 5622, NVN 5627, NVN 5636 and NEN 6421. Gross-alpha en gross-beta activity in solids, filters and water. RIVM report 610013001/2006. (In Dutch).

ENVIRONMENTAL MEASUREMENTS OF RADIOXENON

T.W. Bowyer, J.C. Hayes, J.I. McIntyre

Pacific Northwest National Laboratory, P.O. Box 999, Richland, Washington, USA

1 INTRODUCTION

Radioactive xenon (radioxenon) is produced by the fissioning of nuclear material, either via neutron-induced or spontaneous fission, and also via neutron activation or other nuclear reactions involving xenon gas. The most abundant radioactive xenon isotopes in the atmosphere are ^{131m}Xe, ^{133}Xe and ^{135}Xe, having been measured at several locations in the northern hemisphere associated with reactor operation, medical isotope production, and associated with the spontaneous fission of ^{240}Pu from the legacy materials at plutonium production facility in Hanford, Washington. Radioactive xenon measurement at levels near the average atmospheric level (1-10 mBq/m^3) is a “specialty” measurement, requiring specialized collection, separation, and nuclear measurement techniques primarily because of its low concentration and interference with Rn isotopes. This paper describes the reasons for making radioxenon measurements, background sources, and current techniques for these measurements.

2 HISTORY

Radioactive xenon has been associated with nuclear reactor operation ever since the first reactor “piles” started operation in 1944. Popular accounts of events at the first Hanford piles tell of an incident early on in the project in which a nuclear reaction could not be sustained for more than a few days. Both Enrico Fermi and John Wheeler have been credited with determining the cause of the problem: buildup of a relatively short-lived radionuclide with a very high thermal neutron absorption cross section was poisoning the nuclear reaction by robbing enough of the neutron flux so that sustained nuclear reactions were not possible – the chain reactions were “poisoned.” Data showed that the unknown radionuclide built up after the pile shutdown and then decayed again to levels that allowed the reactor to be started again after about 36 hours. The radionuclide that was ultimately identified as the poison was ^{135}Xe. Xenon-135 has a 9.1-hr half-life and has a huge thermal neutron absorption cross-section of approximately 2×10^6 barns.

The xenon poisoning effect is well known in the field of nuclear engineering as the effect that prevented early reactors from rapid startup after shutdown. In addition to ^{135}Xe being created in high relative yields as an independent yield fission product in the fission of U and Pu, it is also created in high amounts by a chain-yield fission product from decay of ^{135}Te and ^{135}I via:

$$^{135}\text{Te (11 sec)} \rightarrow {}^{135}\text{I (6.7 hr)} \rightarrow {}^{135}\text{Xe (9.1 hr)}.$$

Radioxenon has been measured in reactors routinely to monitor various aspects of their operation. Radioxenon has also been used extensively as a medical isotope. Xenon-133 is primarily used for performing lung diagnostics[1] and measuring blood flow through bodily injection of a saline or other solution in which the radioxenon has been dissolved, and it has also been used for determining tissue characteristics by differential adsorption.[2] Among the attractive aspects of ^{133}Xe is that it is rapidly exhaled after injection or inhalation, and hence does not cause a significant background to other nuclear isotopes used[3] or exposure to patients. In studies involving inhalation of xenon, activity levels as high as approximately 10^9 Bq per treatment are used, causing dose rates to the patient of only approximately 0.2 mSv (the total dose from natural sources is approximately 3 mSv/yr).[4,5,6]

Measurement of radioxenon for the detection nuclear processes and nuclear material has seen significant activity in the last 10 years largely because of technological developments driven by the need to monitor the Comprehensive Nuclear-Test-Ban Treaty (CTBT). During that time, several countries have developed technologies for the measurement of radioxenon and some have been commercialized in the case of automated equipment to detect radioxenon from nuclear explosions.[7,8,9,10]

3 RADIOXENON PRODUCTION MECHANISM

The fissioning of uranium with thermal neutrons produces a characteristic bimodal yield distribution of mass numbers peaking at approximately M=90 and M=140. Xenon has isotopes that are near the maximum of one of the fission mass yield curves, and so relatively large amounts of xenon are released from fission. Due to their chemical nature, noble gases such as xenon do not chemically combine and so they are easily released into the environment, making radioactive xenon a rather unique signature of nuclear processes. In addition to fission, radioxenon isotopes can be produced in other processes such as the irradiation of gaseous xenon or in accelerator-driven reactions. These production mechanisms are not likely a source of large amounts of radioactive xenon in the environment, however, compared to releases from reactors and medical isotope production.

A number of stable and radioactive xenon isotopes exist and are shown in Table 1 along with their natural abundance and cumulative fission yield. The most studied xenon isotopes from the environmental radioactivity perspective are those that both are produced significantly in processes that can release these into the environment and have relatively long half-lives.

Most of the xenon isotopes that have a combination of high fission production cross section and are long-lived enough to be detected (^{131m}Xe, ^{133}Xe, ^{133m}Xe and ^{135}Xe), still have half lives short enough that the airborne activity concentrations measured vary greatly with the distance from and the release rate from the source. For example, although ^{135}Xe is produced in comparable quantities to ^{133}Xe from nuclear reactors, it is less commonly measured in the environment far from reactors, because of its short half-life (9.1 hr versus 5.2 d). Although stable xenon isotopes are also produced in abundance in fission, remote detection of these isotopes above the natural abundance (approximately 87 parts per billion) has not been established, though some efforts are underway to detect small changes in the xenon natural abundance due to fission near nuclear processes.[11]

Table 1. Isotopes of Xenon with half-lives greater than 1 hour.[12]

Isotope	Half-life	^{235}U Thermal Cumulative Fission Yield[1]	Natural Abundance
Xenon-122	20.1 h	<0.01%	
Xenon-123	2.08 h	<0.01%	
Xenon-124	Stable	<0.01%	0.10%
Xenon-125	16.9 h	<0.01%	
Xenon-126	Stable	<0.01%	0.09%
Xenon-127	36.41 d	<0.01%	
Xenon-128	Stable	<0.01%	1.91%
Xenon-129	Stable	0.72%	26.4%
Xenon-129m	8.89 d	<0.01%	
Xenon-130	Stable	<0.01%	4.1%
Xenon-131	Stable	2.9%	21.2%
Xenon-131m	11.9 d	0.03%	
Xenon-132	Stable	4.3%	26.9%
Xenon-133	5.245 d	6.7%	
Xenon-133m	2.19 d	0.19%	
Xenon-134	Stable	7.8%	10.4%
Xenon-135	9.10 h	6.5%	
Xenon-136	Stable	6.3%	8.9%

The processes that emit large amounts of radioxenon into the atmosphere are reactor operation and accidents, medical isotope production and usage, and past nuclear explosions. All other processes are thought to release much lower amounts because of low production yields or as in the case of the reprocessing of nuclear fuel, the radioxenons have decayed away at the time of reprocessing. Since radioxenon produced from past nuclear testing has decayed away, the main sources of radioxenon in the environment are from nuclear reactors and as a by-product of the production of isotopes for medical procedures. Airborne activity concentrations of ^{131m}Xe, ^{133}Xe, ^{133m}Xe, and ^{135}Xe have been reported by several groups[13,14,15,16] over the last few decades and range from a few mBq/m^3 in the reactor-dense regions of the northern hemisphere to as high as several Bq/m^3 close to reactors.[17]

4 XENON COLLECTION, SEPARATION AND MEASUREMENT TECHNIQUES

Physical rather than chemical techniques are commonly used to collect and separate noble gases from air, water, and each other. In large-scale industrial applications, noble gases are nearly exclusively collected from the atmosphere via cryogenic distillation of liquefied air. Perhaps the most common technique for collection and separation of noble gases in smaller scale R&D, however, is through the use of adsorbers such as activated charcoal or molecular sieves, at varying ranges of adsorbent temperature and pressure.[13] Gas chromatography is almost exclusively used for R&D applications to remove contaminants such as N_2, O_2, Ar, and Rn, though other physical separation techniques such as

[1] See for example, http://atom.kaeri.re.kr/

aerodynamic separation processes may possibly be employed to separate Xe from other lower-mass gases.

After it has been collected, radioactive xenon gas can be measured using a variety of techniques including gamma-ray spectroscopy and a technique developed at the Pacific Northwest National Laboratory using beta-gamma coincidence spectrometry.[18] The beta-gamma coincidence spectrometry method allows for the simultaneous measurement of all of the important xenon isotopes with extremely high selectivity,[9] and is the principle detection technique used in the Automated Radioxenon Sampler-Analyzer (ARSA) system designed at PNNL.

5 SOURCES OF RADIOXENON IN THE ENVIRONMENT

5.1 Reactor Accidents

Xe-133 was the main radionuclide released during nuclear incidents such as the Chernobyl and Three Mile Island nuclear accidents. In these accidents, the ^{133}Xe that was built up as a fission product was released. Because of the unreactive nature of the noble gases, the ^{133}Xe was hard to contain and hence was released into the environment.

On April 26, 1986, one of the 1000-MW nuclear power plants in Chernobyl had a devastating accident that released over 10^{18} Bq of radioactivity into the environment. By far, the highest release was from ^{133}Xe, of which at least 2×10^{18} Bq (100% of the noble gas inventory and incidentally the equivalent radioxenon of a 200-kiloton nuclear detonation) was released.[19] Elevated concentrations of ^{133}Xe at levels as high as 400 Bq/m^3 were detected in Europe[20] and as far away as the United States in noble gas monitoring systems in California, Nevada, and Utah.[21]

During the accident at the 850 MW Three Mile Island-2 nuclear reactor on March 28, 1979, an amount of ^{133}Xe comparable to the Chernobyl incident was released. It is estimated that approximately 10^{17} Bq of radioxenon was released when fuel rods partially melted and radioactive noble gases vented into the atmosphere and into the reactor building through a pressure relief valve that stuck open.[22] Elevated ^{133}Xe levels were detected as far away as 375 km in Albany, NY where a peak atmospheric concentration of approximately 144 Bq/m^3 was measured.[23]

Although the Three Mile Island incident of the 1970s and the Chernobyl disaster increased the awareness of radioactive gases in the atmosphere somewhat, radioactive xenon does not pose a significant health risk compared to other radioisotopes since it does not react with the environment and has a short half-life, hence it does not give the public much dose compared to other reactive radioisotopes such as ^{137}Cs and ^{90}Sr even though it is present in the air.

5.2 Reactor Operation and Medical Isotope Production

Radioxenon production in reactors is thought to arise from two pathways, first via xenon emitted from cracks in fuel rods, and secondly from fission of uranium on the exterior of fuel rods or in cooling water. Radioxenon released from medical isotope production comes primarily from the production of ^{99}Mo, in which uranium targets are irradiated in a reactor and ^{99}Mo extracted shortly after the irradiation by chemical dissolution of the targets. Extremely large amounts of radioxenon are released in this process.

Figure 1 shows a plot of the concentration of ^{133}Xe present in the northern hemisphere in central Europe. It has been reported[17] that the concentration at a location in the southern hemisphere (Tahiti) typically shows no measurable radioxenon whatsoever. The difference between the two location levels can be explained by the relative lack of nuclear reactors and medical isotope production facilities in the southern hemisphere. The relative contributions of nuclear reactors and medical isotope production leading to the observed radioxenon levels has not been determined in most cases, but investigations are underway by the International Noble Gas Expert (INGE) collaboration.[8]

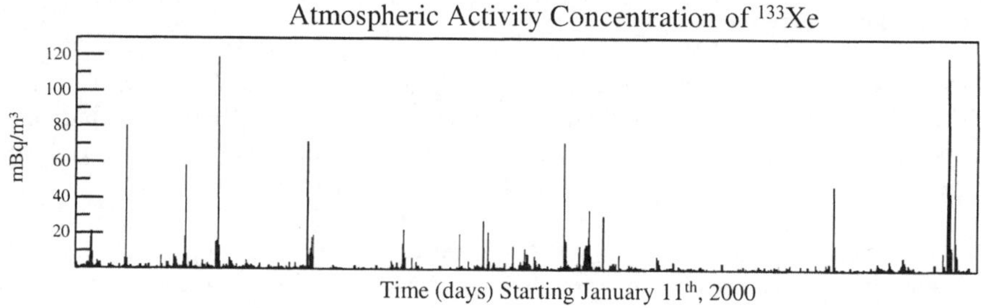

Figure 1 Automated measured atmospheric activity concentrations of ^{133}Xe in Freiburg, Germany for year 2000. The measurement extends over more than 400 days, starting January 11, 2000, and ending on February 28, 2001.[8]

5.3 Nuclear Explosions

Interest has increased in radioactive xenon monitoring significantly since the formation of the International Monitoring System (IMS) for the CTBT in the 1990s. Background information on the concentrations, variability, and isotopes found is needed on a global scale in order for the IMS to assure that the sources of radioxenon can be determined and nuclear explosions can be singled out. At its inception, the IMS lacked data characterizing radioactive xenon on a global scale at levels of approximately 1 mBq/m^3 (and still does), but progress is currently being made.[8]

An issue that is being addressed in the context of global xenon monitoring for nuclear explosions, when the source is not *a priori* known as in the case of local reactor monitoring or subsurface monitoring, is the attribution of an unknown source of radioxenon – namely, what is the source of the measurement? By isotopic analysis of the xenon isotopes in combination authors have suggested that xenon emitted from nuclear explosions can be separated from reactor output since the time it takes to vent species from a nuclear explosion and from a nuclear reactor differ somewhat.[24,25]

5.4 Spontaneous Fission of Nuclear Waste

Radioactive xenon was also expected and in fact observed in measurements of the spontaneous fission of ^{240}Pu in subsurface soils. In this case, the origin of ^{240}Pu in the subsurface was due to Pu entrained in waste discharged into the soil from the United States' weapons program. In this case, the amount of ^{240}Pu in the subsurface soil was estimated to be 10^{-2} to 10^{-1} Ci/g of soil (approximately 0.05 to 0.5 g ^{240}Pu/m^3 soil). This amount created relatively large concentrations of ^{133}Xe and ^{135}Xe in equilibrium.[26]

Radioactive xenon is also produced in spontaneous fission from ^{238}U in soil. However, the most sensitive measurements of ^{133}Xe no better than approximately 0.1 mBq/m^3, and it is not uncommon to measure a value consistent with that level or lower. The concentration of the radioactive xenon isotopes is lower than can be detected in some locations using existing techniques – a rough estimate of the airborne activity concentration of ^{133}Xe that originates from spontaneous fission of ^{238}U can be made by comparing to observed airborne ^{222}Rn levels. Radon-222 concentrations vary from approximately 0 to as high as 100 Bq/m^3, depending on atmospheric conditions as well as local geology.

$$A_{^{133}Xe} = A_{^{222}Rn} \times \left(\frac{\lambda_{S.F.}}{\lambda_{\alpha}}\right) \times \left(\frac{\lambda_{^{222}Rn}}{\lambda_{^{133}Xe}}\right) \times BR_{^{133}Xe} \times \xi$$

A_i = the activity concentration for each isotope,

$\lambda_{S.F.}$ = the spontaneous fission decay constant for $^{238}U = 8.7 \times 10^{-17}\ yr^{-1}$

λ_{α} = the alpha decay constant for $^{238}U = 1.6 \times 10^{-10}\ yr^{-1}$

$\lambda_{^{133}Xe}$ = the decay constant for $^{133}Xe = 0.13\ day^{-1}$

$\lambda_{^{222}Rn}$ = the decay constant for $^{222}Rn = 0.18\ day^{-1}$

$BR_{^{133}Xe}$ = the cumulative yield branching ratio for thermal fission of ^{238}U leading to ^{133}Xe = 0.067

ξ = barometric pumping factor ratio = 1.

The barometric pumping factor ξ, which was set to 1 for this calculation, is the fractional difference in the total amount of Rn and Xe gases pumped out of the ground, ending up in the atmosphere. In reality this factor can vary significantly based on local geology. Taking 100 Bq/m^3 as a maximum average ^{222}Rn concentration, the corresponding activity of ^{133}Xe would be on the order of 1 μBq/m^3, which is much lower than the lowest detectable limits. Therefore, we expect that the ^{133}Xe being produced in the ground via ^{238}U spontaneous fission causes an unobservably low airborne concentration, using common techniques.

5.5 Spent Nuclear Fuel

The concentration of radioxenons in spent fuel is expected to be measurable. For example, spontaneous fission of low burn-up nuclear fuel yielding approximately 0.5 kg ^{240}Pu alone (accompanying 8 kg of ^{239}Pu at 7% burn-up) gives approximately 10^4 Bq of ^{133}Xe at equilibrium. Other contributing isotopes that undergo spontaneous fission (and some neutron-induced fission) in spent fuel, such as ^{238}Pu, ^{242}Pu, ^{242}Cm, ^{244}Cm, will make the radioxenon level even higher. Although there are most likely measurable levels of ^{133}Xe and ^{135}Xe and other radioxenon gases at the time of dissolution of the fuel from spontaneous fission, the signals are far lower than those expected from other gases, such as ^{85}Kr.[27] However, all of these signals including ^{85}Kr are highly diluted in stack gas and are subject to atmospheric transport that make reliable detection of these noble gases impractical at locations outside of the stack of a reprocessing facility.

7 CONCLUDING REMARKS

A significant amount of work has gone into understanding the sources of radioxenon in the environment, as well as how to efficiently, selectively, and sensitively measure the isotopes. Major efforts are continuing to be made to better understand the global distribution of these isotopes, to learn how to attribute elevated levels to a specific source more precisely, and to develop better instrumentation for the measurement of the xenon isotopes.

References

[1] E.R. Powsner and D.E. Raeside, *Diag. Nucl. Med.*, 1971, 519.

[2] L.E. Preuss, F.P. Bolin, D.G. Piper, and C.K. Bugenis, *Int. J. Appl. Radiat. Isot.*, 1975, **26**, 329.

[3] T.J. Herold, M.K. Dewanjee, and H.W. Wahner, *J. Nucl. Med. Technol.*, 1985, **13**, 72.

[4] B. Shleien, L.A. Slaback, and B.K. Briky, in *Handbook of Health Physics and Radiological Health*, 3rd Edn., Williams and Wilkins, Baltimore, 1998, p. 1-5.

[5] P.J. Pityn, M.E. King, and M.J. Chamberlain, *Nucl. Med. Comm.*, 1993, **14**, 1079.

[6] B.F. Peterman, and C.J. Perkins, *Radiat. Prot. Dosim.*, 1988, **22**, 5.

[7] Y.V. Dubasov, Y.S. Popov, V.V. Prelovskii, A.Y. Donets, N.M. Kazarinov, V.V. Mishurinskii, V.Y. Popov, Y.M. Rykov and N.V. Skirda, *Instr. Experim. Techn.*, 2005, **48**, 373.

[8] M. Auer, A. Axelsson, X. Blanchard, T.W. Bowyer, G. Brachet, I. Bulowski, Y. Dubasov, K. Elmgren, J.P. Fontaine, W. Harms, J.C. Hayes, T.R. Heimbigner, J.I. McIntyre, M.E. Panisko, Y. Popov, A. Ringbom, H. Sartorius, S. Schmid, J. Schulze, C. Schlosser, T. Taffary, W. Weiss, and B. Wernsperger, *Appl. Radiat. Isot.*, 2004, **60**, 863.

[9] T.W. Bowyer, C. Schlosser, K.H. Abel, M. Auer, J.C. Hayes, T.R. Heimbigner, J.I. McIntyre, M.E. Panisko, P.L. Reeder, H. Sartorius, J. Schulze, and W. Weiss, *J. Environ. Radioact.,* 2002, **59**, 139.

[10] J.P. Fontaine, F. Pointurier, X. Blanchard, T. Taffary, *J. Environ. Radioact.,* 2004; **72**, 129.

[11] Y. Aregbe, K. Mayer, S. Valkiers, P. De Bièvre, *Fresenius' J. of Anal. Chem.,*1997, **358**, 533.

[12] E. Browne, and R.B. Firestone, in *Table of Radioactive Isotopes*, 1986.

[13] T.W. Bowyer, K,H. Abel, W.K. Hensley, M.E. Panisko, and R.W. Perkins, *J. Environ Radioact.*, 1997, **37**, 143.

[14] C. Kunz, *Atmos. Environ.*, 1989, **23**, 1827.

[15] K. Mueck, *Radiat. Prot. Dosim.*, 1988, **22**, 219.

[16] A. Ringbom, T. Larson, A. Axelsson, K. Elmgren, C. Johansson, *Nucl. Instr. Meth. in Phys. Res. A*, **508**, 2003, 542.

[17] T.J. Stocki, X. Blanchard, R.D'Amours, R.K. Ungar, J.P. Fontaine, M. Sohier, M. Bean, T. Taffary, J. Racine, B.L. Tracy, G. Brachet, M. Jean, and D. Meyerhof, J. Environ. Radioact., 2005, 80, 305.

[18] P.L. Reeder, T.W. Bowyer, and R.W. Perkins, 1998, *J. Radioanal. Nucl. Chem.*, **235**, 89.

[19] M. Eisenbud, *Environ. Radioact.,* 3rd Edn. Academic Press, San Diego, California. 1987, p. 367.

[20] T. Florkowski, J. Grabczak, T. Kuc, and K. Rozanski, 1987, Institute of Nuclear Technology, Krakow, Poland Report 215/I.

[21] R.W. Holloway and C.K. Liu, *Environ. Sci. Technol.*, 1988, **22**, 583.

[22] S. Langer, M.L. Russell, and D.W. Akers, *Nucl. Technol.*, 1989, **87**, 196.

[23] M. Wahlen, C.O. Kunz, J.M. Matuszek, W.E. Mahoney, and R.C. Thompson, *Science* 1980, **207**, 639.

[24] T.W. Bowyer, R.W. Perkins, K.H. Abel, W.K., Hensley, C.W. Hubbard, A.D. McKinnon, M.E. Panisko, P.L. Reeder, R.C. Thompson, and R.A. Warner, *The Encyclopedia of Environmental Analysis and Remediation*, 1998, John Wiley & Sons, Inc., New York, p. 5299.

[25] M. B. Kalinowski, Lawrence H. Erickson, Gregory J. Gugle, Arms Control, Disarmament, and International Security Report, ACDIS KAL:1.2005, 2005.

[26] P.E. Dresel and S.R. Waichler, Pacific Northwest National Laboratory report PNNL-14617, 2004.

[27] M.B. Kalinowski, H. Sartorius, S. Uhl, W. Weiss, *J. Environ. Radioact.,* 2004, **73**, 203.

UPTAKE OF URANIUM BY SPINACH GROWN IN ANDOSOLS ACCUMULATING TRACE AMOUNTS OF FERTILISER-DERIVED URANIUM

N. Yamaguchi,[1] Y. Watanabe,[2] A. Kawasaki[1], C. Inoue[1]

[1] National Institute for Agro-Environmental Sciences, 3-1-3, Kan-nondai, Tsukuba, Ibaraki 305-8604, Japan
[2] Japan Atomic Energy Agency, Tokai Research and Development Centre, 2-4, Shirakata Shirane, Tokai, Naka, Ibaraki 319-1195, Japan

1 INTRODUCTION

Long-term application of phosphate fertiliser results in the elevation of uranium (U) concentrations of agricultural soils as phosphate fertilisers contain appreciable concentrations of U as an impurity[1)-4)]. Water soluble fraction of U accounts for 10 to 80% of total U in phosphate fertiliser. The water-soluble U in the phosphate fertiliser is easily adsorbed on soil[4)], thereby causing an accumulation of U in soil. During a 10-year period in a paddy field of Japan, the increase of U in soil due to the application of phosphate fertiliser was estimated as 5.3% of total U in the soil[4)]. Fertiliser-derived U is mainly associated with organic substances and metal oxides in Andosols, which have higher amounts of organic matter[3)] and could become a potential source for plant uptake following the release of U due to the decomposition of organic matter and/or the dissolution of metal oxides. AnańYan[5)] observed increased U concentrations in meadow grasses due to the application of superphosphates to organic-rich soil.

The World Health Organisation (WHO) established a tolerable daily intake (TDI) for U of 0.6 μg/kg body weight per day[6)]. Direct ingestion or inhalation of dust particles is considered the most critical pathway of radionuclides to humans. In addition, soil-plant-man is recognised as one of the major pathways for the transfer of U to human beings. In order to estimate the intake of U through plants, it is necessary to investigate how U is taken up by food plants.

Uranium absorption by plants has been investigated mainly through small-scale cultivation tests such as pot or lysimeter experiments in which high amounts of U, up to several hundred mg kg^{-1}, are added or mining-impacted soil is used. These approaches have been applied because the absorption of U by plants is lower than that of other trace and hazardous elements in soil[7)], except for some hyperaccumulator plants such as Indian mustard (*Brassica juncea* L.) and sunflower species (*Helianthus annuus* L.)[8)]. In addition, most U is trapped by roots, and only trace amounts transfer to aboveground parts of plants[8), 9)]. The plant/soil concentration ratio (CR) has been used to describe accumulation of radionuclides from soil to plant. Leaf vegetables tended to have higher CR values than rootcrops had, and cereals tended to have lower CR values[10)-13)]. The extrapolation of CR from plants grown in soil with a high U concentration to those grown in soils of near-background U concentrations in agricultural fields is inadequate because of the nonlinearity feature of observed CR. Sheppard and Evenden[11] claimed that CR tended to be lower when the soil concentration was higher. The CR values obtained from small-scale experiments were six-fold higher than those obtained from field-scale experiments. In addition, the U concentration in soil is only one of the factors that influences plant uptake in a field[14]. Different soil types, growth conditions, and climate conditions can also affect CR.

We investigated U absorption by food plants grown in a field. Spinach was chosen because it absorbs relatively higher amounts of U than do other vegetables and crops[10)-13)] The first purpose of this study was to show a field-scale variation range of U concentrations in spinach grown in the same soil but under different growth conditions. The plant physiological reason that spinach can absorb relatively high amounts of U is unknown. We assume that a strategy to absorb Fe from soil of neutral-to-alkaline condition may expedite U availability in rhizosphere soil, therefore enabling spinach roots to absorb U from soil. The second purpose of this study was to find relationships between the uptakes of U and Fe, as well as those of other macro- and microelements in order to find similarities between the uptake of U and other elements.

2 METHODS AND RESULTS

2.1 Uranium in soil for spinach cultivation

Soil samples were taken from the experimental field in which spinach is cultivated at the National Institute for Agro-environmental Sciences, Tsukuba, Japan. The top 10 cm of soil was collected from 15 points and then mixed, air-dried, and passed through a 2-mm sieve. For comparison, surface soil was also taken from the adjacent non-cultivated land. The soil classification based on the world reference base for soil resources is Andosols. Relevant soil properties are shown in Table 1. A distinctive feature of this soil is a high amount of organic matter and a higher capacity for phosphate adsorption due to the presence of non-crystalline minerals.

Table 1 *Soil properties*

pH(H_2O)	pH(KCl)	Total C	Total N	Cation exchange capacity	Exchangeable cations				Particle size distribution (wt%)			
					Ca	Mg	K	Na	Coarse sand	Fine sand	Silt	Clay
		($g\ kg^{-1}$)		($cmol_c\ kg^{-1}$)					0.2 ~ - 2 mm	20 μm ~ 0.2 mm	2 ~ 20 μm	< 2 μm
6.91	5.95	49	4.0	28.7	18.6	4.30	2.20	0.10	6.7	27.2	36.0	30.1

Total, exchangeable, and acid-soluble U concentrations in soil were determined. To determine total concentrations of U in soil, 0.2 g of air-dried soil was digested with HNO_3, $HClO_4$, and HF mixtures using a microwave digester (Multiwave3000, PerkinElmer). After digestion was completed, the acid solution was evaporated to near dryness and then adjusted to 50 mL using 0.16 mol L^{-1} HNO_3. Exchangeable and acid-soluble fractions of U were extracted by 1 mol L^{-1} CH_3COONH_4 and 0.44 mol L^{-1} CH_3COOH for 24 h at a soil-to-solution ratio of 1 to 10. The U concentration was determined using an inductively coupled plasma mass spectrometer (ICP-MS, Platform, MicroMass) equipped with a flow-injection system. Table 2 shows the concentrations of total, exchangeable, and acid-soluble U in the soil. The U concentration in 1 mol L^{-1} CH_3COONH_4 extract from the soil was lower than the background U concentration of the extractant. The exchangeable U fraction in soil was negligibly low in both the cultivated and non-cultivated fields, suggesting that U availability to plants was low. The concentration of acid-soluble and total U in the cultivated field soil was higher than that in the adjacent non-cultivated surface soil, which suggested that agricultural practices caused increased U concentrations in ploughed soil. Uranium impurities in the phosphorus fertiliser were the most probable source of the increased U concentrations.

Table 2 *Uranium concentrations in soils (mean ± standard uncertainty).*

	Total U	Exchangeable U	Acid-soluble U	
		(mg kg^{-1})		% of Total U
Spinach field soil	2.1 ± 0.06	< 0.0001	0.035 ± 0.001	1.6
Non-cultivated soil	1.1 ± 0.10	< 0.0001	0.0093 ± 0.0005	0.84

2.2 Uranium concentrations in spinach

Three cultivars of spinach were grown on the above-mentioned field where the long-term application of phosphorus fertiliser may have caused elevated concentrations of U in ploughed soil. Before seeding, 40 m^2 of the field was fertilised with 5.5 kg of high-analysis compound fertiliser (N:P:K = 13-7.9-11.6), 7.2 kg of dolomitic limestone, 20 kg of bark compost, 20 kg of poultry manure, and 20 kg of humus compost. Spinach was harvested after 36 - 62 days (*Spinacia oleracea* L. cv. Action and *S. oleracea* L. cv. Active) or 126 - 171 days (*S.* L. cv. Solomon) of growth, when the leaf rosettes were fully developed but before flower stalk shot up (Table 3). A longer cultivation period was required for cv. Solomon because it was grown in winter. After harvest, the above-ground parts of the plants were separated into leaves and petioles. They were carefully washed in tap water until all visible soil particles on the tissues were removed, and then they were rinsed by shaking in ultra-pure water for five repetitions. The washed spinach samples were freeze-dried and powdered by hand in a polyethylene bag pre-washed with ultra-pure water. The concentration of U was determined by ICP-MS after microwave-assisted digestion of 0.2 g of powdered sample with HNO_3 and H_2O_2. The drift of the instrument response and matrix-induced signal suppression were compensated for the addition of an internal standard (Bi 1 μg L^{-1}).

Table 3 compares the effects of washing methods on U concentrations in selected leaf samples. Concentrations of U in the leaf samples before the additional washing with ultra-pure water are shown in the parentheses. The U concentrations before washing with ultra-pure water were significantly higher than those after washing based on a paired sample's t-test at a 0.05 level of significance. Since U is ubiquitous in soil, adhering soil particles can contribute to the U concentration in spinach if the washing treatment is insufficient. In this study, the effects of adhering soil particles on U concentrations in spinach tissues were minimised.

The concentrations of U on a dry-weight basis were higher in leaves than in petioles, ranging from 0.67 to 5.92 μg kg^{-1} in leaves and from 0.24 to 2.37 μg kg^{-1} in petioles (Table 3). Even though the spinaches were grown in the same field soils, the U concentrations varied with different conditions, such as various temperatures and amounts of precipitation, during the cultivation period. No clear relationships were found among the U concentrations, days of cultivation, and accumulated amounts of precipitation. In addition, among the three spinach cultivars used in this study, no clear differences in U concentrations were found.

Table 3 *Uranium concentrations in spinach and relevant growth conditions (mean ± standard uncertainty).*

Cultivars	Days of cultivation	Accumulated precipitation (mm)	U concentration ($\mu g\ kgDW^{-1}$)	
			Leaf	Petiole
Active	36	148	1.87 ± 0.03	0.572 ± 0.040
Active	37	97	3.60 ± 0.14 (6.89 ± 0.44)*	1.13 ± 0.03
Active	41	119	0.923 ± 0.03	0.560 ± 0.040
Active	53	115	1.48 ± 0.03	0.686 ± 0.092
Active	62	270	1.72 ± 0.05	0.526 ± 0.010
Action	37	210	1.76 ± 0.07 (3.92 ± 0.32)*	0.463 ± 0.030
Action	39	69	0.671 ± 0.07 (1.36 ± 0.09)*	0.236 ± 0.054
Action	45	159	1.81 ± 0.49	0.880 ± 0.166
Action	47	247	5.92 ± 0.08	2.37 ± 0.13
Solomon	126	291	3.52 ± 0.18	1.99 ± 0.14
Solomon	171	425	2.64 ± 0.11	0.825 ± 0.015
Average			2.36 ± 1.50	0.930 ± 0.666

*Numbers in parentheses show U concentrations in spinach leaf before being washed by ultra-pure water.

For comparison, U concentrations in some selected agricultural products grown in the same field were also determined and are listed in Table 4 in order of increasing U concentration. Skins of the root crops were removed in order to avoid contamination by soil particles. However, except for spinach and Ching Guang Juai, samples were not additionally washed in ultra-pure water. Mean U concentrations in edible parts of leaf vegetables including spinach were higher than those of cereals and rootcrops as suggested in previous reports[10-13]. Uranium concentrations in the above-ground parts of spinach cultivated in 13 cycles ranged from 0.60 to 8.49 $\mu g\ kg^{-1}$ dry weight and from 0.037 to 0.61 $\mu g\ kg^{-1}$ fresh weight. The maximum concentration was observed for cv. Action grown in summer. For this sample, U concentrations in leaves and petioles were not determined and are not shown in Table 2 because of the low yield. High-temperature injury could have occurred; however, it was unclear whether the injury caused a higher U uptake. The calculated CR values for spinach on a dry- and fresh-weight basis ranged from 2.8×10^{-4} to 4.0×10^{-3} and from 1.8×10^{-5} to 2.9×10^{-4}, respectively. The CR values for spinach reported by Sasaki et al.[12] were similar : 4.7×10^{-3} on a dry-weight basis and 3.6×10^{-4} on a fresh-weight basis. The CRs for mixed vegetables reported by IAEA[15] were also of the order of 10^{-4} to 10^{-3}. On the other hand, higher CR values of the order of 10^{-2} for spinach grown in mining-impacted soil were reported (0.87 and 0.74 $\mu g\ g^{-1}$ in spinach and 17 $mg\ kg^{-1}$ in soil; U concentrations in spinach based on whether dry or fresh weight were not reported[16]). As in the case of spinach, U concentrations in other plants should vary widely depending on growth and climate conditions. Nevertheless, apart from the divergence of the CR values, the amounts of U taken up by plants grown in Andosols reported in this study tended to be lower than previously reported[10)-13)]. The amounts of radionuclides taken up by plants were controlled not only by the total concentration of the soil but also by the availability of radionuclides in the soil. The lower CR values for plants grown in Andosols would be caused by the high adsorption capacity of Andosols which is responsible for the lower availability of U to plants.

Table 4 *Uranium concentrations in edible parts of plants grown in the same field.*

Common name	Scientific name	*n*	U concentration in edible part (μg kgDW^{-1})
Horse bean	*Vicia faba* L.	1	0.08
Soybean	*Glycine max* L.	1	0.10
Pumpkin	*Cucurbita moschata* (Duchesne ex Lam.) Duchesne ex Poir.	1	0.12
Potato	*Solanum tuberosum* L.	2	0.13
Tomato	*Lycopersicon esculentum* Mill.	1	0.19
Upland rice	*Oryza sativa* L.	1	0.22
Sweet potato	*Ipomoea batatas* L. Poir.	1	0.25
Shallot	*Allium chinense* G.Don	1	0.33
Snow pea	*Pisum sativum* L. var. macrocarpum Ser.	1	0.34
Adzuki bean	*Vigna angularis* Ohwi et H.Ohashi var. angularis	1	0.35
Cauliflower	*Brassica oleracea* L. var. botrytis L.	1	0.39
Okra	*Abelmoschus esculentus* L. Moench	1	0.46
Eggplant	*Solanum melongena* L.	1	0.48
Broccoli	*Brassica oleracea* L. var. italica Plenck	1	0.48
Welsh onion	*Allium fistulosum* L.	1	0.52
Taro	*Colocasia esculenta* L. Schott	1	0.70
Japanese radish	*Raphanus sativus* L. var. hortensis Backer	1	0.72
Chinese cabbage	*Brassica rapa* L. var. glabra Regel 'Pe-tsai'	1	0.73
Butter bean	*Phaseolus vulgaris* L.	1	0.84
Cucumber	*Cucumis sativus* L.	1	0.98
Corn	*Zea mays* L.	1	1.1
Buckwheat	*Fagopyrum esculentum* Moench	1	1.2
Cabbage	*Brassica oleracea* L. var. capitata L.	2	1.3
Ching Guang Juai	*Brassica rapa* L. var. chinensis	4	1.7±0.8
Bell pepper	*Capsicum annuum* L. 'Grossum'	1	1.8
Lettuce	*Lactuca sativa* L.	1	2.3
Carrot	*Daucus carota* L. subsp. sativus (Hoffm.) Arcang.	1	2.5
Spinach	***Spinacia oleracea* L.**	13	**3.1±2.4**
Komatsuna	*Brassica rapa* L. var. perviridis L.H.Bailey	1	5.4
Garland chrysanthemum	*Glebionis coronaria* (L.) Cass. ex Spach	1	5.5
Chard	*Beta vulgaris* L. subsp. cicla (L.) Koch var. cicla L.	1	6.8
Mulukhiya	*Corchorus olitorius* L.	1	9.5
Ginger	*Zingiber officinale* Roscoe	1	9.6

2.3 Relationships among concentrations of U and other trace elements.

Spinach absorbed higher amounts of U than other vegetables and crops did, in spite of the fact that the U availability in the soil was low. The uptake of U by spinach from soil can be accompanied by the uptake of other elements that are also insoluble in soils. In order to compare U and other trace elements taken up by spinach tissues, concentrations of Mg, Al, Ca, Fe, Cu and Zn in acid digests of spinach samples were analysed by inductively coupled plasma optical emission spectrometry (VistaPro, Varian).

Figures 1 and 2 show relationships among concentrations of U and selected major and trace elements in spinach leaves and petioles, respectively. It is noteworthy that concentrations of U in spinach were significantly positively correlated (p<0.01) with concentrations of Fe and Al in both leaves and petioles. These relationships suggested that the absorption and transport processes of U in spinach could be related to those of Fe and Al, as was also suggested by Kametani et al.[17)], who showed that plants with higher Fe concentrations tended to absorb more U. Less U was extracted by 1 mol L^{-1} ammonium acetate solution from soil (Table 2), meaning that U in soil was less available to plants. Spinach favours neutral-to-weak alkaline conditions and has the ability to acquire insoluble mineral nutrients such as Fe under neutral-to-alkaline conditions. Helal et al.[18)] compared spinach and beans with respect to the ability of the root to uptake Fe and found that spinach root absorbed Fe more efficiently. The differences in Cu, Zn, and Cd uptake by two spinach cultivars were attributed to different abilities to exude oxalate, citrate, and malate from root[19)]. The application of organic acids to soil facilitated the phytoextraction of U by hyperaccumulator plants[20)]; thus, those root exudates could induce U dissolution from soil. Since part of U is associated with Fe and Al minerals in the soil[3)], it was likely that the absorption of U was accompanied by Fe and Al absorption, possibly triggered by the secretion of protons or organic acids to solubilise Fe and Al from soil.

Because of the physicochemical similarity, some toxic metals are absorbed by plants using the same uptake and transport pathways as those used by essential elements. For example, phytosiderophores can facilitate the absorption of Al into the cytoplasm through the Fe(III)-phytosiderophore transport system[21)] because the ionic radii of Al^{3+} and Fe^{3+} are similar. Metals taken up by plants from roots are transported to aboveground parts through the xylem, and this transport is promoted by the transpiration of water via leaves. In the xylem vessels, Fe is transferred in complexed form with organic acids[22)]. However, the average mole ratios of U to Fe and of U to Al in spinach tissues are extremely low: 4.3×10^{-6} and 3.2×10^{-6} in leaves and 5.0×10^{-6} and 4.5×10^{-6} in petioles, respectively. After being taken up by roots, the ultra-trace level of U could be transported to petioles and leaves along with water flow without passing through the Fe-transport pathway. In order to find plant physiological meanings in terms of the relationships between U and Fe or Al uptakes, further investigations, including studies of the interaction between U and the Fe-transporter, are required.

3 CONCLUSION

Amounts of U absorbed by spinach grown in Andosols under field conditions were 2.36 ± 0.13 μg kg^{-1} in leaves and 0.93 ± 0.05 μg kg^{-1} in petioles. The U concentrations in spinach leaves and petioles were significantly related to Fe and Al concentrations. Uranium uptake by spinach is accompanied by Fe and Al uptake.

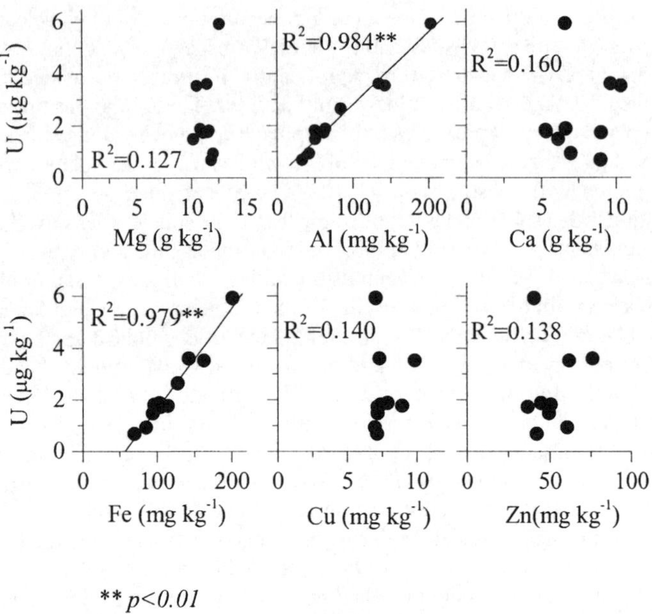

Figure 1 *Relationships between concentrations of U and other elements in spinach leaves.*

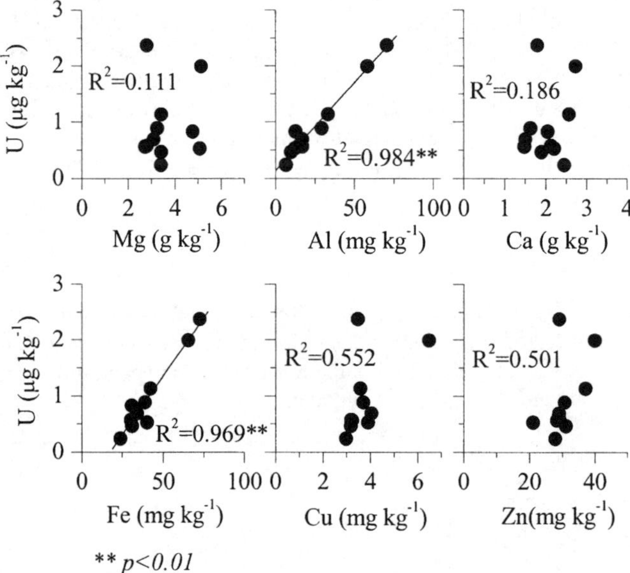

Figure 2 *Relationships between concentrations of U and other elements in spinach petioles.*

References

1 R.F. Spalding and W. M. Sackett, 1972, *Science*, **175**, 629.
2 R.L. Jones, 1992, *Commun. Soil Sci. Plant Anal.,* **23**, 67.
3 A. Takeda et al. 2006, *Sci. Total Environ.,* **367**, 924.
4 A. Tsumura and S. Yamasaki, 1993, *Radioisotopes*, **42**, 265. *in Japanese*
5 V.L. Anańyan, 1989, *Agrokhimiya,* **10**, 100.
6 World Health Organization. 2005, *Uranium in drinking water,* http://www.who.int/water_sanitation_health/en/
7 J.J. Mortvedt, 1994, *J. Environ. Qual.,* **23**, 643.
8 H. Shahandeh and R.Hossner, 2002, *Water, Air, Soil Pollut.,* **141**, 165.
9 A.W. Roberts and M.E. Robb, 1972, *Science,* **178**, 980.
10 S.C. Sheppard, W.G. Evenden, R.J. Pollock, 1988, *Can. J. Soi. Sci.*, **69**, 751.
11 S.C. Sheppard and W.G. Evenden, 1988, *J. Environ. Radioactiv.,* **8**, 255.
12 T. Sasaki, Y. Tashiro, H. Fujinaga, T. Ishii, Y. Gunji, 2002, *Jpn. J. Health phys.,* **37**, 208. *in Japanese*
13 K.P. Singh, 1997, *Current Sci.*, **73**, 532.
14 B.D. Amiro and S.C. Sheppard, 1987, *Health Phys.*, **52**, 233.
15 IAEA, 1994, *Technical reports Series*, **364**, 14.
16 M.R. Saric, M. Stojanovic, M. Babic, 1995, *J. Plant Nutr.,* **18**, 1509.
17 K. Kametani and T. Matsumura, 1993, *Radioisotopes,* **42**, 407. *in Japanese*
18 H.M. Helal, H. Arisha, E. Rietz, 1990, *Plant and Soil,* **123**, 229.
19 H. Keller and W. Romer, 2001, *Plant Nutr. Soil Sci.,* **164**, 335.
20 J.W. Huang, M.J.Blaylock, Y. Kapulnik, B.D. Ensley, 1998, *Environ. Sci. Technol.* **32**, 2004.
21 L.V. Kochian, 1995, *Annu Rev. Plant Physiol. Plant Mol. Biol.,* **46**, 237.
22 M. Greger, 'Metal Availability, Uptake, Transport and Accumulation in Plants' in *Heavy Metal Stress in Plant*, ed. M. N. V. Prasad, Narosa Publishing House, New Delhi, 1999, Chapter 1, p. 1.

MINERALOGICAL AND PARTICLE SIZE CONTROLS ON ^{137}Cs ABUNDANCES IN DOUNREAY OFFSHORE AND FORESHORE SANDS

Ian W. Croudace[1], Phillip E. Warwick[1] and Joe Toole[2]

[1]GAU-Radioanalytical, National Oceanography Centre, European Way, Southampton, SO14 3ZH, UK
[2]Particles Monitoring Manager, UKAEA, Dounreay, Caithness

1 INTRODUCTION

The Dounreay site was established as the site of the UK Fast Breeder Nuclear programme in 1955 and became operational in 1958. It accommodated three reactors, the Materials Test Reactor, (DMTR, 1958-1969), the Dounreay Fast Reactor (DFR, 1959-1977) and the Prototype Fast Reactor (PFR, 1974-1994). With all reactor operations now finished and the reactors already de-fuelled the site is undergoing active decommissioning which is planned to be completed by 2032. The Dounreay site has been cited by UKAEA as being the second biggest nuclear decommissioning challenge in the UK with similar liabilities to those at Sellafield but with smaller waste volumes.

The main aim of this study was to determine the mineral sorption sites for ^{137}Cs in the marine sands from the Dounreay area and in particular for those from an on-site location associated with a seepage from a low-active drain. The interaction mechanism of the discharge with the minerals in the sand was also pertinent to offshore sands, of similar mineralogy, that are exposed to leached fission products from fuel particles which are known to occur as a legacy of the site operations. This legacy arose from the unintentional release of radioactive particles into the environment over an extended period. It is now considered that the releases probably occurred during the sixties and seventies when fuel debris from ponds entered an effluent discharge stream. Millimetric and sub-millimetric particles (sand-sized) of spent nuclear fuel have been discovered and retrieved from local beaches since 1983 and from off-shore sands since 1997. These particles are all strictly catalogued, investigated and stored and UKAEA have spent over £10m to date on scientific surveys and research to understand and ameliorate the 'particles' legacy. So far nearly 240 particles have been found on the Dounreay foreshore and 929 have been recovered from offshore areas. Estimates of numbers of particles released are very imprecise and may never be known. The current study was conceived to understand the interaction of leached and soluble fission products with Dounreay sands (offshore and onshore). Such information could then contribute to the site operator's longer-term risk evaluation for the particles and seep-contaminated sand.

2 METHODOLOGY

The approach used in the study consisted of a sequence of tests on three sands.

a) **DST-MTR:** This Dounreay sand was taken from a core at the end of a 5-month laboratory tank test (Warwick and Croudace 2005). In this experiment small-scale passive breakdown and diffusive transport of radionuclides had occurred from a millimetre-scale fuel fragment (MTR Fuel Particle 02-397-01) lying in the middle of a cylindrical body of seawater-saturated sand.
b) **DST-Raw:** This was an uncontaminated Dounreay sub-tidal /offshore sand that was identical to that used in the tank experiments.

c) **DFS:** This was a Dounreay foreshore sand which consisted of coarse sand which was contaminated with anthropogenic radionuclides from a seepage.

A range of mineralogical and chemical methods were applied to determine the associations of ^{137}Cs with mineralogy and particle size.

2.1 Granulometry

Particle size distributions were determined using a wet-sieving system (Fritsch) to separate the sand mixtures into particle size ranges. It comprised an electromechanical base unit fitted with a sieve stack consisting of seven stainless steel sieves (1400 µm, 710µm, 500µm, 355µm, 250µm, 100µm and 63µm).that separated the mineral grains into eight convenient size intervals. The stack was placed on a base collector having a discharge spout fitted with a silicone tube leading to a 5 litre bucket. The top unit of the sieve stack incorporated a 3 port spray head that allowed a controllable flow of water to pass onto the sand.

2.2 X-Ray Diffractometry

A Panalytical X'Pert PRO X-ray diffractometry system was used to determine bulk mineralogy. Samples were ground using a micronising mill along with added alumina (to act as an internal standard) to facilitate the later quantification of the diffraction data.

2.3 High Resolution gamma spectrometry

A Canberra well-detector system using Fitzpeaks gamma spectrometry software was used to determine caesium-137 activities in all sample fractions. Variations in sample height arising from small sample masses were corrected for where appropriate. The detector efficiency calibration was determined using an NPL mixed gamma standard spiked into a sand matrix (Croudace 1991).

2.4 Magnetic separation of minerals

Mineral grains were separated according to their magnetic susceptibility using a Franz Isodynamic™ magnetic separator. The device was run at full power to effect a separation of non-magnetic minerals (e.g. quartz, feldspars and calcite) from the more magnetic ferromagnesian minerals (e.g. biotite, muscovite, amphiboles, pyroxenes, magnetite). The efficiency of mineral separations into magnetic and non-magnetic portions depended on how extensively mineral grains of contrasting susceptibility were attached or not.

3 RESULTS

The initial results showed that though mineralogically similar (Figure 1 and 2) both sand samples have different particle size distributions which reflects their depositional environment. The sands are composed nominally of sub-equal amounts of quartz and felsdpars (~28 wt%) and carbonate (~40 wt%) with small amounts of mica (5 wt% biotite and muscovite). The foreshore sand is generally coarser than the offshore sand with a maximum mass fraction at 710μm compared with 250μm for the offshore sand. The mineralogy of each particle fraction is approximately similar (Figure 2) except for the <63μm material which is mostly finely-comminuted biogenic carbonate with minor clay minerals. The ^{137}Cs activity is distributed across the range of particle sizes (Figure 3, 4). The results of magnetically separating the >250μm fraction of the DST-Raw and DST-MTR and the >500μm DFS show that the magnetic fraction (Figure 6,7 and 8) is composed of dark minerals found to be a mixture of mostly biotite but also containing some muscovite and minor amphibole. The least magnetic fraction contains mostly quartz, feldspars and calcite.

3.1 Acid leaching and ^{137}Cs uptake experiments

Acid leaching treatments of subsamples were used to establish whether the carbonates held any ^{137}Cs activity and how easily the^{137}Cs activity could be removed from the silicate fraction (Fig 5). The results showed that the biogenic carbonates and the acid leachate had virtually no ^{137}Cs activity; the silicates are therefore the candidate phases that hold the radio-caesium strongly.
The uptake of ^{137}Cs by uncontaminated Dounreay offshore sand was investigated to confirm which mineral were involved in caesium sorption (DST-R; 250μm fraction was tested). The results showed that the more-magnetic mineral fraction had a higher sorption capacity than the non-magnetic fractions. It is notable that the interaction time of 24hr may have been too short to establish maximum uptake potential.

Table 1 *^{137}Cs uptake experiments on DST-Raw 250μm fractions - interaction time of 24 hours*

	Sample	Seawater (g)	Sand (g)	Bq added	Total Bq in filtrate	Bq on solid	Dist. Coeff.
Most Mag	DST-R#4	10.84	0.5807	186	51.67	134.33	49
	DST-R#3	10.77	0.5081	186	112.92	73.08	14
	DST-R#2	10.53	0.3770	187	149.79	37.32	7
Least Mag	DST-R#1	10.85	0.6377	187	159.47	27.63	3
Raw 250; no mag clean	DST-R#0	10.83	0.6252	186	151.54	34.10	4
Second mag clean of DST-R#1	DST-R#5	10.86	0.6108	186	175.70	9.93	1

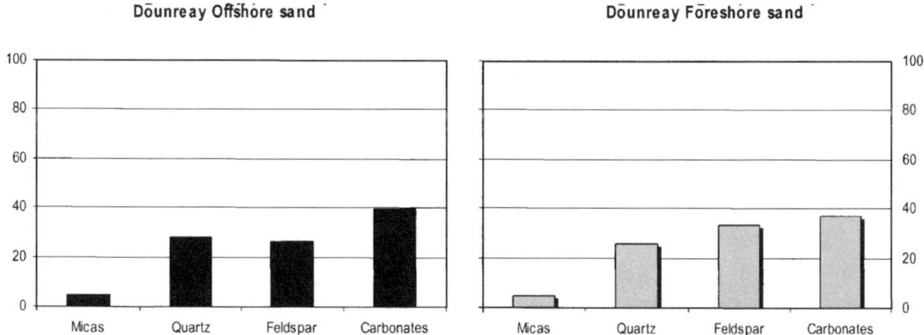

Figure 1. *General bulk mineralogical composition (%) for Dounreay Offshore and Foreshore sand*

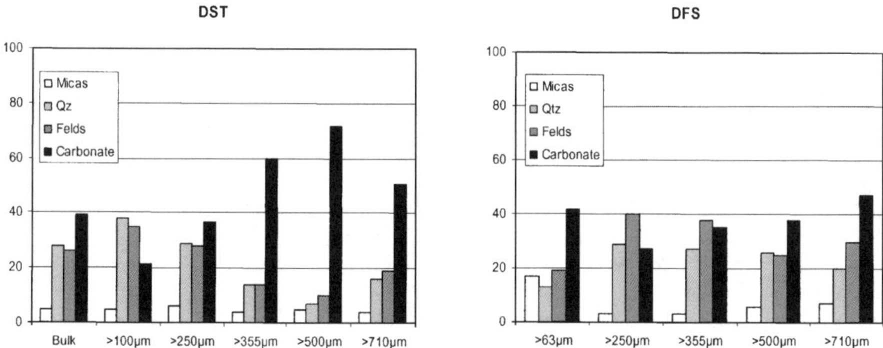

Figure 2. *Variation of mineralogical composition (%) with size fraction for Dounreay Offshore and Foreshore sand*

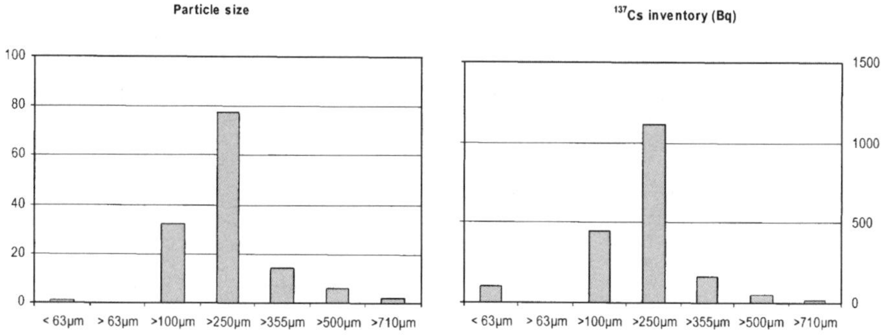

Figure 3. *Particle size distribution (%) and total ^{137}Cs activity (Bq) for Dounreay offshore sand (DST-MTR)*

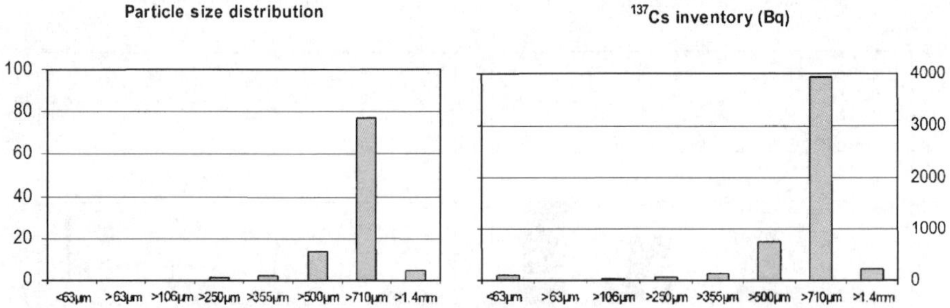

Figure 4. *Particle size distribution (%) and total ^{137}Cs activity (Bq) for Dounreay foreshore sand DFS*

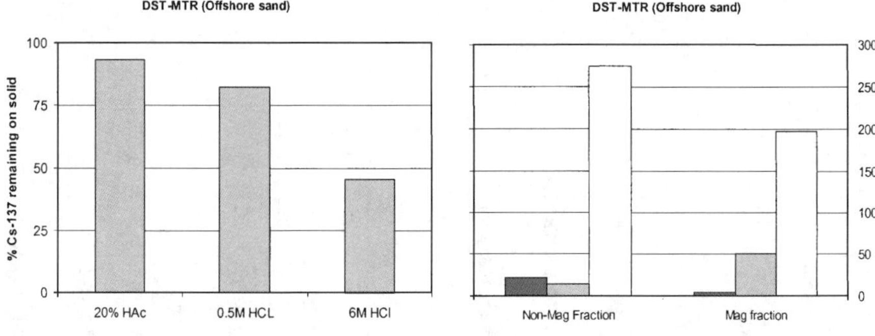

Figure 5. *Effect of acid leaching on >250µm sand fraction DST-MTR*

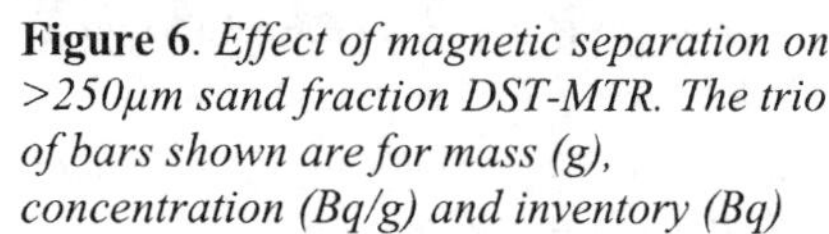

Figure 6. *Effect of magnetic separation on >250µm sand fraction DST-MTR. The trio of bars shown are for mass (g), concentration (Bq/g) and inventory (Bq)*

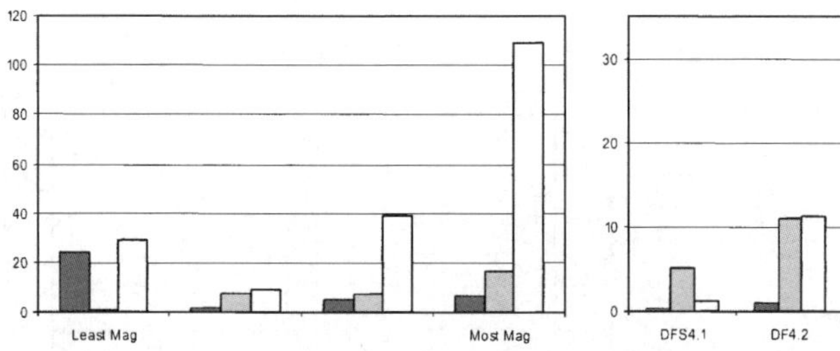

Figure7. *Association of ^{137}Cs mineral fractions for Foreshore sample DFS 500µm fraction (susceptibility increases left to right);. the trio of bars shown are for sample mass (g), ^{137}Cs concentration (Bq/g) and inventory (Bq)*

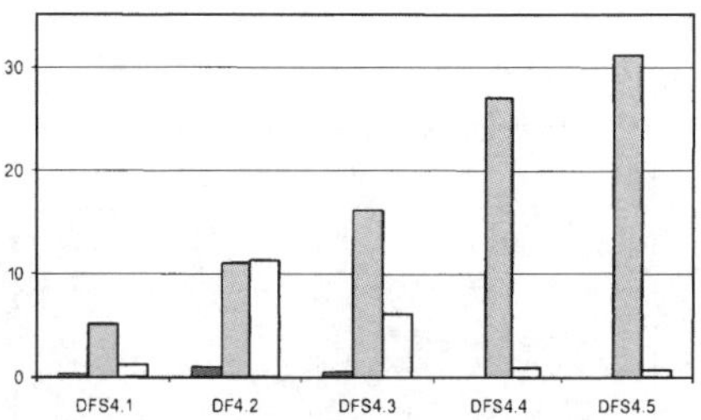

Figure 8. *Effect of further magnetic separation on the most-magnetic fraction of the Foreshore sample DFS#4 (>500µm fraction) (susceptibility increases left to right). The trio of bars shown are for sample mass (g), ^{137}Cs concentration (Bq/g) and inventory (Bq)*

4 DISCUSSION

The various studies described above show that ^{137}Cs is distributed throughout the particle size range in the samples examined. The removal of the appreciable carbonate fraction from the silicates showed very little loss of radiocaesium indicating that it is not carbonate associated. It also shows that ^{137}Cs is not readily exchangeable and is effectively sorbed to or trapped in silicate mineral lattices. The strong positive correlation of ^{137}Cs inventory with the mass of each particle size fraction (Figures 3 and 4) shows that the radionuclide is sorbed to mineral phase(s) present in all size fractions and that each fraction contains approximately the same mineralogy (Figure 2). The use of magnetic separation indicates that ^{137}Cs is associated significantly with the magnetic mineral fraction (Figures 6, 7, 8). The magnetic separation of the sand fractions produced mineral concentrates that are relatively pure. Some impurities of magnetically incompatible minerals occur owing to inherent attachments or static attraction between grains. Gentle crushing and washing followed by repeated magnetic separations would be required to improve purification of the mineral phases. A range of evidence points to the micas as important host minerals for the ^{137}Cs. This includes the petrographic observations (Figure 1 and 4) and the elevated ^{137}Cs inventories in the magnetic fraction. The relative paucity of clay minerals in the Dounreay sands, which might otherwise be likely sorption sites for ^{137}Cs, explains why the micas are the main host for ^{137}Cs in these sands. Approximately 5 wt% of biotite, muscovite and ferromagnesian minerals (e.g. amphibole and pyroxene) occur in the bulk sand and >30% in the magnetic mineral fraction.

The importance of micaceous minerals for sorbing and retaining Cs is well recognised (e.g. Maes *et a*l 1999, Zachara *et al* 2002 , Steefel *et a*l 2003. McKinley *et al.* 2004). It is known that the uptake occurs at two rates, a rapid early uptake followed by a slower sorption following prolonged contact with Cs^+ ions. These micas are susceptible to developing frayed edges where caesium can become non-exchangeably trapped. A schematic representation and empirical X-ray mapping data (Figure 9) from McKinley *et al* (2004) show how Cs can become trapped in micas. Corroboration of this proposed mechanism would require further work.

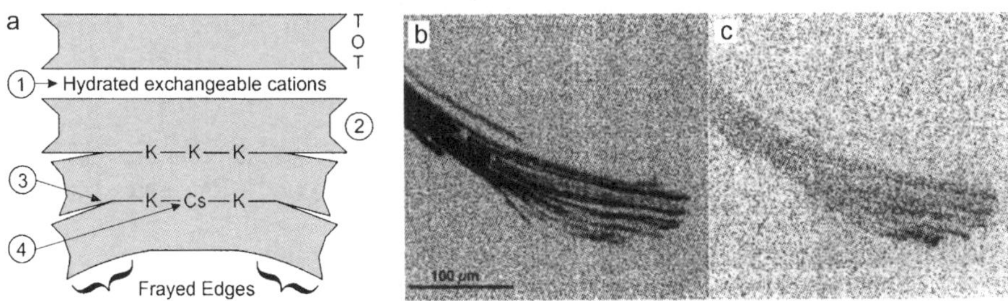

Figure 9: *a) Schematic diagram of the mica structure showing how Cs ions are sorbed at frayed edge sites where K ions have been lost. Possible sites for Cs are 1) cation exchange sites on the mica basal plane, (2) edge sites, (3) frayed edge sites, (4) replacement of K by Cs in interlayer sites. TOT refers to the tetrahedral-octahedral-tetrahedral sheet structure of micas. b) Elemental X-ray maps demonstrating the loss of K from frayed edge sites and c) the occupancy of Cs ion at frayed edge sites (right of image); taken from McKinley et al. 2004*

5 CONCLUSIONS

A range of mineralogical and geochemical investigations were carried out to determine the controls on ^{137}Cs abundances in mineralogically similar sands originating from on-shore and offshore areas near the Dounreay site. The ^{137}Cs from the on-shore sand originated from an on-site seepage whereas the off-shore sand was that used following a 5-month laboratory tank experiment where leaching rates from a fuel particle were being studied. The sands from both locations are composed of quartz (~28 wt%), feldspar (~27 wt%) and carbonates (~40 wt%) with small amounts of mica (5 wt% biotite and muscovite). The small quantity of <63μm material is composed of micas, clays and finely comminuted carbonate and gypsum. The ^{137}Cs activity was found to be distributed across the range of particle sizes. Although elevated concentrations are seen in the <63μm fraction this only represents 0.2 wt% of the sand and therefore was not the main sink for ^{137}Cs. Acid leaching of the sand samples with acetic acid and hydrochloric acid indicated that the ^{137}Cs was not readily leachable with mild acids and is non-exchangeable at least with respect to protons. Following a separation into magnetic and non-magnetic fractions ^{137}Cs was found to be significantly associated with the dark magnetic fraction of the sand which consisted mostly of micas (biotite and muscovite). This association is consistent with reports from elsewhere where it is argued that Cs may become incorporated at frayed-edge sites that develop in altered and weathered micas.

Acknowledgements
This study was funded by UKAEA (Contract 04-01005). We are grateful for the invaluable technical expertise of Trevor Clayton and Ross Williams (X-ray diffraction analysis) and Dr Nicola Holland (optical microscopy).

References

I. W. Croudace (1991) A reliable and accurate procedure for preparing low-activity efficiency calibration standards for germanium gamma-ray spectrometers. *J. Radioanalytical Nuclear Chemistry Letters*, **153,** 151-162.

I. W. Croudace and P. E. Warwick (2005) Mineralogical and particle size controls on ^{137}Cs abundances in Dounreay offshore and foreshore sands. *GAU634 Feb 2006* Commissioned study by UKAEA 04-1005extension.

E. Maes, L. Vielvoye., W. Stone, B. Delvaux, (1999) Fixation of radiocaesium traces in a weathering sequence mica → vermiculite → hydroxy interlayered vermiculite. *European J.Soil Sci,*. **50**, 107-115.

J. P. McKinley, J. M. Zachara, S. M. Heald, A. Dohnalkova., M. G. Newville, S. R. Sutton, (2004) Microscale distribution of cesium sorbed to biotite and muscovite. *Env. Sci. Technol.,* **38**, 1017-1023.

C. I. Steefel, S. Carroll, P. Zhao, S. Roberts (2003) Cesium migration in Hanford sediment: A multisite cation exchange model based on laboratory transport experiments. *J Contamin. Hydrol.*, **67**, 219-246.

P. E. Warwick and I. W. Croudace (2005) Laboratory experiments into loss of radioactivity from fuel fragments *GAU464 Dec 2005* Commissioned study by UKAEA 04-1005.

J. M. Zachara, S. C. Smith, C. Liu, J. P. McKinley, R. J. Serne, P.L. Gassman, (2002) Sorption of Cs+ to micaceous subsurface sediments from the Hanford site, USA. *Geochim. Cosmochim. Acta,* **66**, 193-211.

ASSESSMENT OF POSSIBLE SOURCES OF ARTIFICIAL LONG-LIVED RADIONUCLIDES IN ENVIRONMENTAL SAMPLES BY MEASUREMENT OF ISOTOPIC COMPOSITION

Z. Varga[1], G. Surányi[2], N. Vajda[3] and Z. Stefánka[1]

[1]Institute of Isotopes, Hungarian Academy of Sciences, Konkoly-Thege utca 29-33, H-1121, Budapest, Hungary
[2]Research Group of Geophysics and Environmental Physics, Hungarian Academy of Sciences, Pázmány Péter sétány 1/C., H-1117, Budapest, Hungary
[3]Radiochemistry Laboratory, Institute of Nuclear Techniques, Budapest University of Technology and Economics, H-1521, Budapest, Hungary

1 INTRODUCTION

Long-lived transuranic radionuclides (e.g., ^{237}Np, ^{239}Pu or ^{240}Pu) have been released into the environment as a result of nuclear weapon tests,[1,2] accidents of nuclear power plants, reprocessing plants[3] and nuclear-powered satellites.[4] Plutonium, as an almost exclusively anthropogenic radionuclide, is one of the most radiotoxic actinide elements as a consequence of its high formation yield, dose contribution and long half-life; thus determination of its concentration is of high importance.[5] Americium-241, as it is formed by the decay of ^{241}Pu, is one of those few artificial radionuclides the concentration of which is increasing in the environment, and it is also an important radionuclide from radioecological aspects. Beside the radionuclide concentration the isotopic composition of the sample is also of great interest for the evaluation of the possible origin of contamination, which can be exploited in environmental monitoring, nuclear safeguards and nuclear forensic studies.[6,7] Recently, the use of dispersed plutonium in the environment as a tracer is intensively investigated to study erosion processes and marine transport.[8]

Plutonium has been introduced into the environment through various sources with different isotopic patterns. Origin of environmental plutonium together with the main isotopic characteristics is summarized in Figure 1. The isotope ratios are highly dependant on the exact production circumstances (e.g., burn-up, reactor or weapon design, initial uranium enrichment). Generally, plutonium derived from reactor operation after sufficiently long irradiation and as a consequence of high neutron flux is characterized by relatively high ^{238}Pu content (formed by the neutron capture of ^{237}Np, through the (n,2n) reaction of ^{239}Pu and the decay of ^{242}Cm), high ^{240}Pu/^{239}Pu (up to 0.6 atom ratio depending on reactor-type and burn-up) and ^{241}Pu/^{239}Pu ratios. Moreover, presence of other transuranics (e.g., higher amount of ^{242}Pu, ^{244}Cm) is also characteristic for reactor origin.

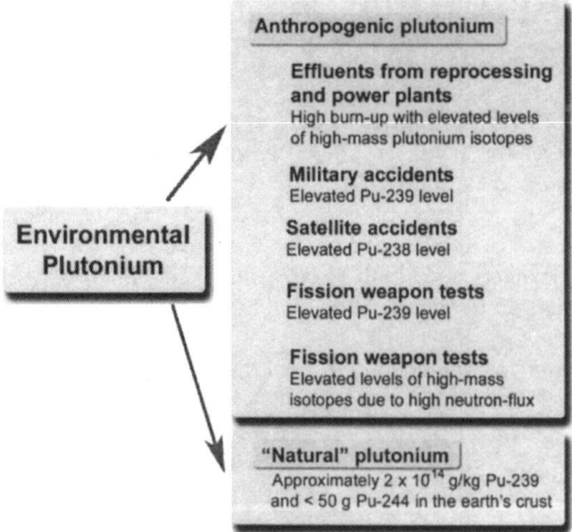

Figure 1 *Origin of environmental plutonium with the main isotopic characteristics*[3,5,9]

Weapons-grade plutonium, dispersed at military accidents such as Thule[9] in 1968 or as non-fissioned weapon particles after detonation of a Pu-bomb can be characterized by high ^{239}Pu content relative to the other Pu-isotopes, while accidentally dispersed Pu from the previously widely used nuclear-powered satellites are characterized by high ^{238}Pu content.[4] The ratio of americium-241 to plutonium isotopes (as ^{241}Am is formed by the decay of ^{241}Pu) is proportional to the initial ^{241}Pu concentration, thus it can also be used as an indicator to assess the origin of contamination. However, in most cases, as several sources may contribute to the transuranics content in environmental samples, mixing models applying several isotope ratios are required to assess the origin of possible contamination sources.

In order to determine plutonium and ^{241}Am isotope ratios for the characterization of environmental samples improved sample preparation methods combined with inductively coupled plasma sector field mass spectrometry (ICP-SFMS) and alpha-spectrometry were developed. The procedures involve a CaF_2 co-precipitation for pre-concentration and matrix removal followed by different extraction chromatographic separation schemes. Samples from contaminated and non-contaminated areas were analysed by both radioanalytical and mass spectrometric techniques. The obtained isotope compositions in the environmental samples were used to assess the sources of contamination. The advantages and limitations of different plutonium and americium isotope ratios (^{238}Pu/^{239}Pu, ^{240}Pu/^{239}Pu, ^{241}Pu/^{239}Pu and ^{241}Am/^{239}Pu) for the characterization of the contaminating sources are discussed.

2 METHOD AND RESULTS

2.1 Investigated Samples

Various sample types (soil, sediment, biota) from different locations were analysed. In most cases commercially available reference materials were used due to their well-documented sampling conditions and wide variety of sampling sites. Furthermore, bioindicators, especially moss samples were measured. These samples were from higher-contaminated (collected by the Hungarian Nuclear Society in Chernobyl and Pripjaty) and less-contaminated (Hungary and South-France) sites. The sample types, sampling locations and dates are summarized in Table 1.

Table 1 *Type, sampling location and date of analysed samples*

Sample name	*Sample type*	*Sampling location*	*Sampling date*
AUSSED	Sediment	Neusiedlersee, 80 km south-east of Vienna, Austria	1986
AUSSOIL	Soil	Ebensee, Upper Austria	1983
BALTSED	Sediment	Baltic Sea	1986 October
CHERMOSS	Moss	Red Forest, Chernobyl Exclusion Zone, Ukraine	2005
ENEWSED	Sediment	Enewetak, Central Pacific Ocean	1990
HUNBIOTA	Biota (mixed weed and grass)	Central Hungary	1986 May
HUNMOSS	Moss	Gerecse Mountains, Hungary	2004
HUNMOSS2	Moss	Gerecse Mountains, Hungary	2004
IRSED	Sediment	Irish Sea	1995
IRSED2	Sediment	Lune Estuary, Irish Sea	1991
MEDWEED	Seaweed	Shore of Principality of Monaco	1986 October
MONMOSS1	Moss	La Turbie, Monaco	2006
MONMOSS2	Moss	Tete de Chien, Monaco	2006
PRIPMOSS	Moss	Pripjaty, Chernobyl Exclusion Zone, Ukraine	2005
FANSED	Sediment	Fangataufa, French Polynesia	1996
RUSSOIL	Soil	Novozybkov, Brjansk, Russia	1990

2.2 Sample Preparation

The applied sample preparation methods used for the analysis are described in details elsewhere.[9,10] In short, after addition of ^{242}Pu and ^{243}Am tracers followed by complete leaching or total dissolution the plutonium as Pu(III) and americium were pre-concentrated using selective CaF_2 co-precipitation step in reductive acidic medium. After dissolving the CaF_2(Am,Pu) precipitate the Pu and Am were separated from the matrix by different extraction chromatographic methods using TEVA, UTEVA and TRU (Eichrom Inc., USA) resins (Figure 2). The main advantage of the above-described pre-concentration step beside its rapidity and simplicity is that it is carried out in acidic media leaving the major matrix components (e.g. alkali metals and most transition metals) and possible interferences in extraction chromatographic separation (e.g. phosphate, sulphate) and mass spectrometric analysis (e.g. uranium) in the liquid phase. The removal of iron as a stable fluoride-complex assures the proper conditions for the TRU separation, while the uranium forms highly soluble hydroxylamine complex. This selective pre-concentration step also overcomes the problem of overloading the extraction chromatographic column. After the pre-concentration three different extraction chromatographic separation methods were applied for the actinide analysis. The first method (Method A) is applicable for Pu analysis by alpha-spectrometry and ICP-SFMS. The analyte is separated applying TEVA column, using dilute HNO_3/HF solution for Pu elution after the removal of U and Th. The second separation method (Method B) that uses UTEVA and TRU columns in tandem arrangement was used for Pu and ^{241}Am analysis by both techniques, while the third separation scheme (Method C) can be applied for ^{241}Am measurement by ICP-SFMS. The final fractions obtained were divided into two parts for alpha-spectrometric and ICP-SFMS analysis. The alpha sources were prepared using NdF_3 micro co-precipitation.

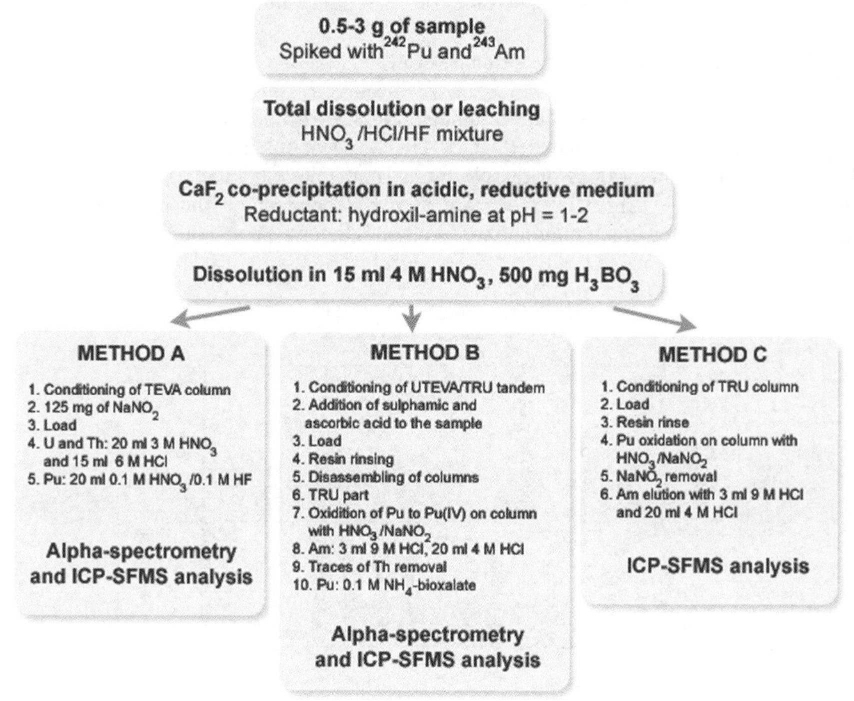

Figure 2 *Flow-chart of the applied sample preparation methods*

2.3 ICP-SFMS and Alpha Spectrometric Analysis

The ICP-SFMS instrument was tuned using a 1 ng ml^{-1} uranium standard solution prior to analysis. Sensitivity was about 2 x 10^6 cps for 1 ng g^{-1} ^{238}U solution. Concentrations of plutonium isotopes and ^{241}Am were calculated as a function of ^{239}Pu/^{242}Pu, ^{240}Pu/^{242}Pu, ^{241}Pu/^{242}Pu and ^{241}Am/^{243}Am ratios according to the isotope dilution method. All raw data were corrected for instrumental mass bias using linear correction.[11] NdF_3 micro co-precipitated alpha sources were counted by a PIPS type alpha Si detector with a surface area of 450 mm^2 attached to an alpha spectrometer (Canberra Inc., USA). Counting time was typically 2-5 days, the detector background was less than 10^{-5} cps in the applied 3.5-9 MeV energy range. The efficiency of the detector was 0.32-0.41 depending on geometry. The overall uncertainty was calculated according to error propagation rules taking into account the uncertainties of weight measurements, tracer concentration and the measured intensities. For ^{239}Pu, ^{240}Pu, ^{241}Pu and ^{241}Am limits of detection of 2.4 fg g^{-1} (0.006 mBq), 1.5 fg g^{-1} (0.013 mBq), 2.0 fg g^{-1} (7.7 mBq) and 1.0 fg g^{-1} (0.13 mBq) were achieved by ICP-SFMS, respectively, and 0.02 mBq by alpha-spectrometry for ^{238}Pu, $^{239+240}$Pu and ^{241}Am. The decontamination factors of the interfering nuclides (e.g., ^{228}Th or ^{210}Po in case of alpha-spectrometry and ^{238}U, Pb, Bi for ICP-SFMS) were 10^3-10^6, which was sufficiently high to obtain pure measurement sources. The methods were validated by the analysis of reference materials and the results were also checked by the comparison of ICP-SFMS ^{239}Pu+^{240}Pu and alpha spectrometric $^{239+240}$Pu results.

2.4 Isotopic Patterns of the Investigated Samples

The isotopic ratios of the investigated samples are shown in Figure 3-6. Typical isotope ratios of reactor-grade plutonium (Reactor GP[3,9]), weapons-grade plutonium (Weapon GP[12]) and global fallout values of the Northern Hemisphere[4,5,13] are also indicated. Note that ^{241}Pu/^{239}Pu and ^{241}Am/^{239}Pu ratios are highly dependant on the decay time, while isotope ratios in reactor-grade plutonium depend on reactor-type and burn-up.[9]

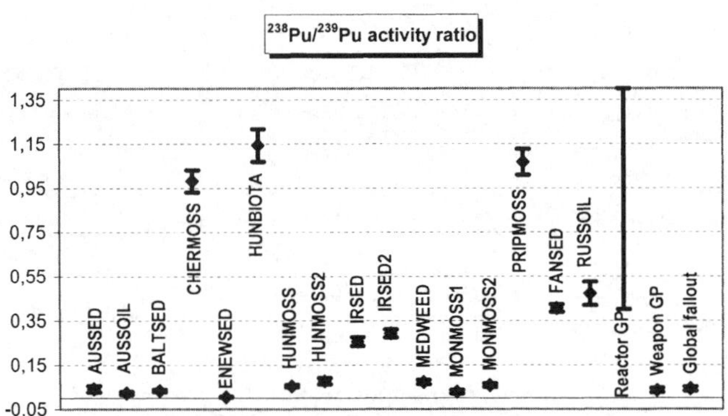

Figure 3 The *^{238}Pu/^{239}Pu activity ratios in the analysed samples*

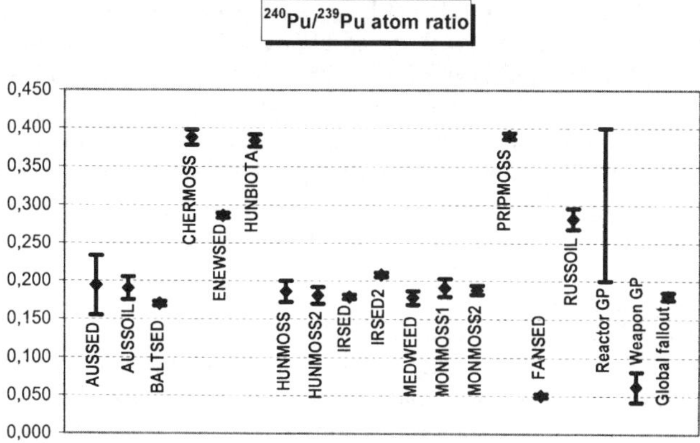

Figure 4 *The $^{240}Pu/^{239}Pu$ atom ratios in the analysed samples*

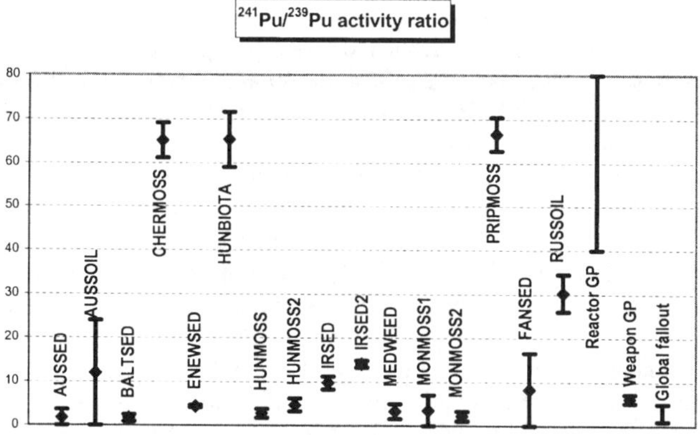

Figure 5 *The $^{241}Pu/^{239}Pu$ activity ratios in the analysed samples*

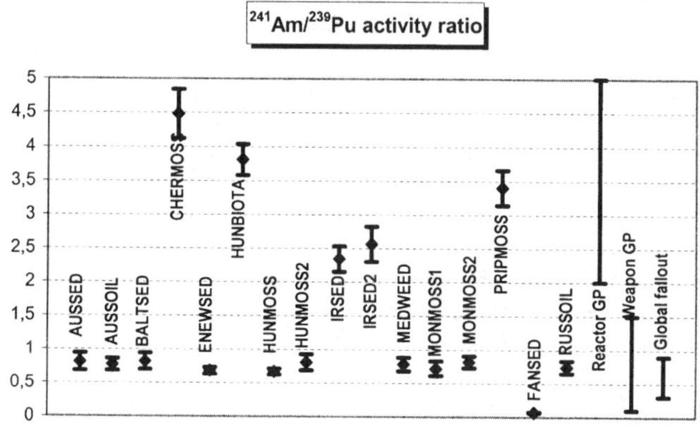

Figure 6 *The $^{241}Am/^{239}Pu$ activity ratios in the analysed samples*

In most of the samples collected at less-contaminated sites the measured isotope ratios do not differ significantly from those of global fallout. However, in the case of IRSED, IRSED2, RUSSOIL and HUNBIOTA samples higher values were obtained. This can be explained by possible contamination from Sellafield reprocessing plant in the Irish Sea sediment samples[6] and from Chernobyl in the case of Russian soil sample.[3] The HUNBIOTA sample (collected in Hungary right after the Chernobyl accident) has Pu and ^{241}Am isotope ratios almost identical to those of Chernobyl and Pripjaty samples, which confirms that most of contamination in these samples derives from the Chernobyl accident. By the results it can be established that ^{238}Pu/^{239}Pu activity ratio is a more sensitive indicator for distinguishing global fallout from reactor-grade plutonium than ^{240}Pu/^{239}Pu and ^{241}Pu/^{239}Pu ratios. This derives from the fact that ^{238}Pu/^{239}Pu ratio varies over a wider range and is not prone to the decay of the shorter-lived ^{241}Pu ($T_{1/2}$ = 14.35 years). Moreover, as ^{240}Pu/^{239}Pu atom ratio differs only by a factor of two for global fallout and reactor-grade plutonium isotope ratios (approximately 0.18 and 0.4, respectively), small changes in the ratios are masked by the relatively low ICP-SFMS measurement precision (between 5-15% in case of low-level plutonium analysis). Instruments with better precision (e.g., thermal ionization mass spectrometry or multi-collector ICP-SFMS) can be applied instead. To overcome the problems due to decay of previously dispersed ^{241}Pu, ^{241}Am/^{239}Pu ratio is preferred for the assessment of origin. Nonetheless, because of the different chemical properties and bioavailability of Pu and Am the results have to be carefully evaluated. Moreover, because the ^{241}Am/^{239}Pu ratio is also highly dependant on decay time, it can be used to calculate the age of the incident, if initial ^{241}Am/^{239}Pu is known.

In the case of multiple contamination sources, mixing models can be applied for the evaluation.[14,15] Using a two-component model, i.e., assuming that plutonium and americium derive from two sources with known isotope ratios, the contribution of the different sources, e.g., in the Irish Sea sediment samples and the Russian soil, can be assessed. For the calculation ^{238}Pu/^{239}Pu and ^{240}Pu/^{239}Pu ratios were used as the most sensitive indicator. However, similar results may be obtained by the other isotope ratios, but with higher uncertainty. In the case of Irish Sea samples, IRSED and IRSED2, knowing that Sellafield release had a ^{238}Pu/($^{239+240}$Pu) ratio of 0.1-0.3 depending on the date of release,[16] the contribution of Sellafield to the total plutonium is between 42% and 86%. The highest uncertainty in this calculation is the initial isotope ratio of Sellafield release that varies with a factor of two, as available literature data is highly unreliable. In the case of RUSSOIL and HUNBIOTA samples, when it is assumed that plutonium derives from the Chernobyl accident with well-defined and known isotope ratios[3] beside global fallout, results with lower uncertainty may be obtained. For the RUSSOIL and HUNBIOTA samples the calculated Chernobyl-contribution values are 44 ± 4% and above 95%, respectively.

Samples from contaminated areas have significantly different isotopic composition than the typical global fallout values. The isotope ratios of Chernobyl (CHERMOSS) and Pripjaty (PRIPMOSS) moss samples agree with previously reported values.[3,5,14] Beside the high ^{238}Pu, ^{240}Pu and ^{241}Pu values, significant amount of $^{243+244}$Cm was also observed in the alpha spectra indicating high burn-up and reactor origin. Samples from previous weapon-test sites (ENEWSED and FANSED) show different isotopic composition. In the case of Enewetak sample the very low ^{238}Pu/^{239}Pu activity ratio and relatively high ^{240}Pu/^{239}Pu ratio are similar to those found after the Chinese fusion device test (H-bomb) in 1976 in Lop Nor by De Geer.[17] The elevated level of high-mass Pu isotopes is explained by the multiple captures of neutrons in the natural uranium blanket in the extremely high neutron fluencies. In Fangataufa sample the high ^{238}Pu content may be formed by (n,2n) reaction (average fissioning energy is about 10 MeV, while the threshold energy of this reaction is about 6 MeV),[17] whilst low ^{240}Pu/^{239}Pu ratio agrees with that of the weapons-grade plutonium[12] and presumably the plutonium contamination derives from a previous fission test (^{239}Pu-bomb).

3 CONCLUSION

Improved rapid sample preparation methods were developed for plutonium and ^{241}Am analysis by alpha-spectrometry and ICP-SFMS techniques. The methods were applied for the analysis of various samples from highly contaminated and less-contaminated sites. It was demonstrated that using different isotope ratios obtained both by mass spectrometry and alpha-spectrometry as complementary tools the source of plutonium contamination can be assessed and retrospective investigation is possible.

4 ACKNOWLEDGEMENTS

Part of this study was carried out within a Technical Cooperation with the International Atomic Energy Agency at the Marine Anvironment Laboratory, Monaco and financially supported by the Hungarian Atomic Energy Authority (OAH-ANI-ABA-05/06) and the Hungarian Scientific Research Fund (OTKA F61087). Dr. Tamas Biro (Institute of Isotopes) is thanked for the helpful discussion.

References

1 M. Yamamoto, A. Tsumura, Y. Katayama and T. Tsukatani, *Radiochim. Acta,* 1996, **72**, 209.

2 M.B. Cooper, P.A. Burns, B.L. Tracy, M.J. Wilks, and G.A. Williams, *J. Radioanal. Nucl. Chem.* 1994, **177**, 161.

3 *The International Chernobyl Project*, Technical Report, Assessment of Radiological Consequences and Evaluation of Protective Measures. Report by an International Advisory Committee, IAEA, Vienna, 1991.

4 E.P. Hardy, P.W. Krey and H.L. Volchok, *Nature*, 1973, **241**, 444.

5 *Ionizing Radiation: Sources and Biological Effects*, United Nations Scientific Committee on the Effects of Atomic Radiation, United Nations, New York 1982.

6 J.S. Becker, *Int. J. Mass Spectrom.* 2005, **242**, 183.

7 M. Betti, G. Tamborini and L. Koch, *Anal. Chem.* 1999, **71**, 2616.

8 S.H. Lee, J. Gastaud, J.J. La Rosa, L.L.W. Kwong, P.P. Povinec, E. Wyse, L.K. Fifield, P.A. Hausladen, L.M. Di Tada and G.M. Santos, *J. Radioanal. Nucl. Chem.* 2001, **248**, 757.

9 Z. Varga, G. Surányi, N. Vajda and Z. Stefánka, *Radiochim. Acta,* (accepted).

10 Z. Varga, G. Surányi, N. Vajda and Z. Stefánka, *J. Radioanal. Nucl. Chem.* (accepted).

11 K.G. Heumann, S.M. Gallus, G. Rädlinger and J. Vogl, *J. Anal. At. Spectrom.* 1998, **13**, 1001.

12 M.J. MICHOLAS, K.L. COOP, R.J. ESTEP, *Capability and Limitation Study of DDT Passive-Active Neutron Waste Assay Instrument*, Los Alamos National Laboratory, LA-12237-MS, 1992.

13 T. Warneke, I.W. Croudace, P.E. Warwick and R.N. Taylor, *Earth Planet. Sci. Lett.* 2002, **203**, 1047.
14 M.E. Ketterer, K.M. Hafer and J.W. Mietelski, *J. Environ. Radioact.* 2004, **73**, 183.
15 J.M. Kelley, L.A. Bond and T.M. Beasley, *Sci. Total Environ.* 1999, **237/238**, 483.
16 J. Merino, J.A. Sanchez-Cabeza, L. Pujol, K. Leonard and D. McCubbin, *J. Radioanal. Nucl. Chem.* 2000, **243**, 517.
17 L.-E. De Geer, 'The Radioactive Signature of the Hydrogen Bomb' in *Science & Global Security*, Gordon and Breach Science Publishers S.A., 1991, Vol. 2, pp. 351-363.

A RAPID METHOD FOR THE PRECONCENTRATION OF NON-REFRACTORY Am AND Pu FROM 100 g SOIL SAMPLES

E. Philip Horwitz[1]*, Anil H. Thakkar[2], and Daniel R. McAlister[1]

[1]PG Research Foundation, Inc.
[2]Eichrom Technologies, Inc.

1 INTRODUCTION

During the last few years several studies have been published dealing with actinide analysis from 2 to 10 g soil and sediment samples[1-4]. To avoid the influence of different matrices, these techniques involved a preconcentration step to remove the major ions from the soils and sediments. After preconcentration, the actinides are then separated from any remaining matrices and from each other using extraction chromatographic techniques[5,6]. Although these methods have given consistent chemical recoveries (~85%), they are not rapid and are not easily adaptable to large (100 g) soil samples.

We have developed a novel rapid method for the preconcentration of non-refractory transuranics (such as fallout) from 100 gram soil samples. Experiments with reference soils suggest the acid-leachable fraction most likely represents the potentially available transuranic elements and, therefore, is of potential value. The basis of our technique is the exploitation of a substantial synergistic extraction of Am(III) from 1 to 6 M HCl in the presence of ferric chloride by a commercially available tetra-n-octyldiglycolamide (referred to hereinafter as DGA) extraction chromatographic resin. Although the uptake of Am and Pu from pure HCl by the DGA resin can be moderately high[6], it is insufficient (especially in the case of Am) to efficiently remove these transuranic elements from a large volume of acidic HCl solutions containing complex matrices. The uptake of Pu(IV) is also enhanced by the presence of large amounts of ferric chloride but less than Am(III). However, Pu(IV) is sufficiently strongly retained on DGA resin from HCl without the need for $FeCl_3$. Since Fe is usually a major constituent in soils, a small DGA column is sufficient to retain Am and Pu from a large volume of 3 to 4 M HCl even in the presence of high concentrations of Fe(III) that are in excess of the capacity of the resin. Figure 1 shows the acid dependency of k' (the number of free column volumes to peak maximum) of Am(III) versus HCl using DGA resin in the absence of Fe(III) and in presence of three different concentrations of iron. The k' was measured as described previously[7].

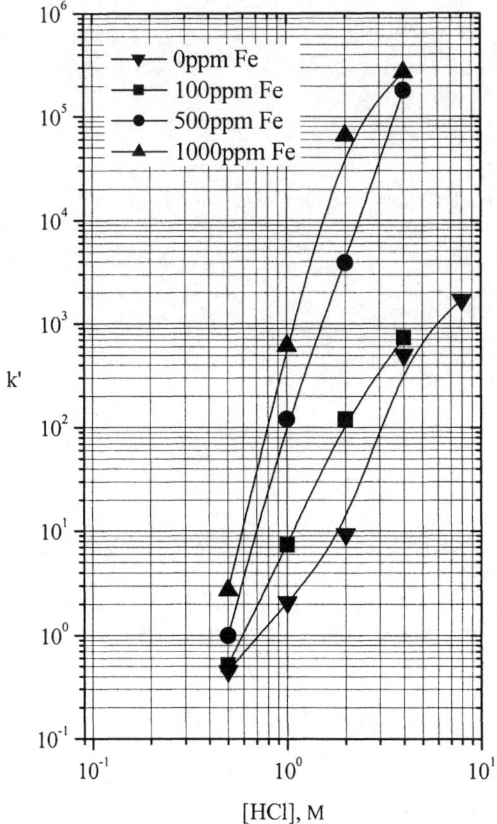

Figure 1. *Acid dependency of k' for Am from HCl + Fe on DGA (Fe added as $FeCl_3{\cdot}6H_2O$)*

2 EXPERIMENTAL

2.1 General Preconcentration Strategy

Our overall approach to matrix elimination and Am and Pu preconcentration is shown in Figure 2. One hundred grams of a dried Northern Illinois soil sample is muffled at 500°C for 3 hours in a glass beaker. Four hundred μg of Sm(III) and the appropriate amounts of Pu(III)/(IV) and Am(III) tracers are added to the muffled soil sample. The sample is then leached with 300 mL of 6 M HCl at 90°C for 2 hours with stirring. After cooling, the supernatant is separated by centrifugation for 15 minutes at 5000 rpm. The residue is washed two successive times with 75 mL of 6 M HCl and the combined supernatant plus rinses (~450 mL) is filtered to eliminate any suspended particles.

The filtered leachate solution (~3.5 M in HCl) is loaded onto a 1.5 cm i.d. column containing 0.625 grams of DGA resin (50-100 μm), which gives ~2 mL bed volume. (The DGA resin is obtained from Eichrom Technologies, Inc., Darien, IL 60561, USA.) The column is preconditioned with 5 mL of 4 M HCl prior to use.

After loading, the column is rinsed twice with 5 mL of 4 M HCl, (the first rinse is from the beaker containing the load solution) and twice with 2.5 mL of 3 M HNO_3. The Am and Pu are then stripped with 20 mL of 0.25 M HCl – 0.03 M oxalic acid. Gravity flow rate (~7 to 8 mL/min) is used throughout the column run.

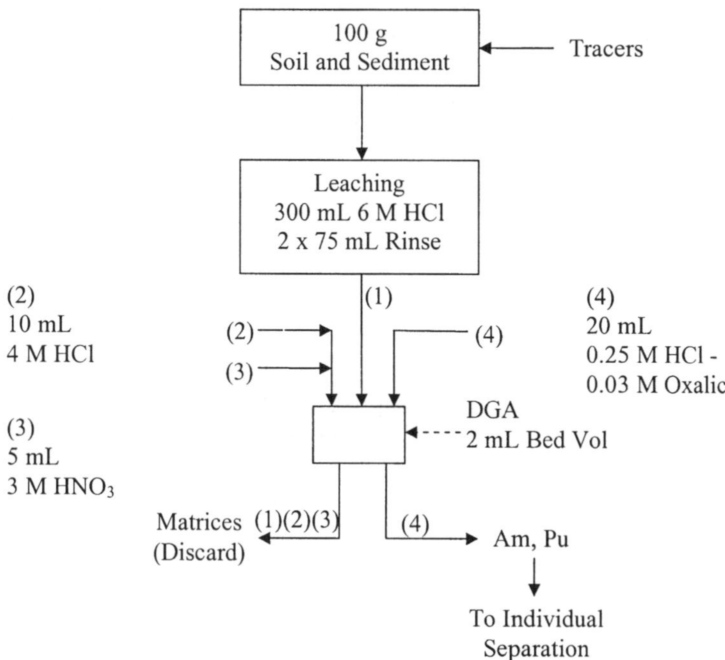

Figure 2. *Flowchart for the preconcentration of Am and Pu from 100 gram soil sample*

2.2 Actinide Separation

After evaporation to dryness in the presence of H_2O_2, the strip solution from the DGA column is compatible with published extraction chromatographic procedures for isolating Am and Pu. However, in the interest of time, we avoid evaporation and peroxide addition at this stage by acidifying the strip solution to 3 M HNO_3 with 4.7 mL of concentrated HNO_3, giving a final volume of 25 mL. Figure 3 shows a flowchart of the separation scheme employed. A key step involves the coupling of 2 ml TEVA and DGA cartridges. (The TEVA and DGA cartridges were obtained from Eichrom Technologies, Inc.) The DGA cartridge is attached to the bottom of the TEVA cartridge and then placed in a vacuum box (also from Eichrom). The tandem cartridges are preconditioned with 5 mL of 3 M HNO_3 using a flow rate of 1 mL/min. The load solution is mixed with 2.3 mL of 3.5 M $NaNO_2$ (to ensure Pu is in the tetravalent oxidation state) and then passed through the tandem TEVA and DGA cartridges at a flow rate of 1 mL/min. After rinsing the beaker plus column with 10 ml of 3 M HNO_3 at a flow rate of 3 mL/min, the cartridges are separated for individual treatment. The TEVA cartridge is eluted successively with 10 ml of 3 M HNO_3, 20 ml of 9 M HCl (to remove Th) and 20 mL of 0.1 M HCl/ 0.05 M HF/0.03 M $TiCl_3$(Pu fraction). The Pu fraction is now radiochemically pure and is ready for source preparation for alpha counting. Alpha sources are prepared by CeF_3 coprecipitation and filtration[8]. Fifty μg of cerium nitrate, 0.5 mL of H_2O_2 and 5 mL of concentrated HF are added in succession to the Pu fraction. The solution is then allowed to stand for 30 minutes before filtering using Eichrom Resolve® filters.

The separated DGA cartridge is rinsed with 5 mL of 3 M HNO_3 and 20 mL of 0.1 M HNO_3. Americium is then stripped with 20 mL of 0.1 M HCl. The Am fraction is evaporated to dryness, redissolved in 10 mL of 4 M ammonium thiocyanate/0.1 M formic acid and loaded onto a 2 mL TEVA cartridge that is preconditioned with 5 mL of the thiocyanate/formic acid reagent. After rinsing the TEVA cartridge with 10 mL of 1.5 M ammonium thiocyanate/0.1 M formic acid (to remove lanthanides), the Am is eluted with 25 mL of 0.25 M HCl. The Am fraction is now radiochemically pure and is ready for source preparation as described above for the Pu fraction except that the H_2O_2 is eliminated.

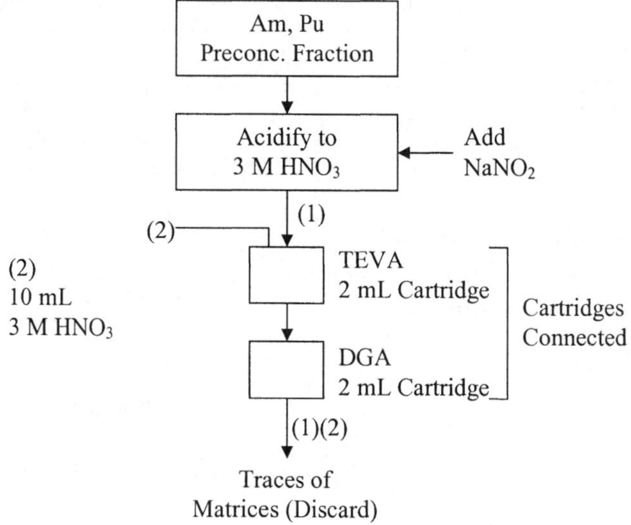

Figure 3. *Flowchart for front-end of individual actinide separations*

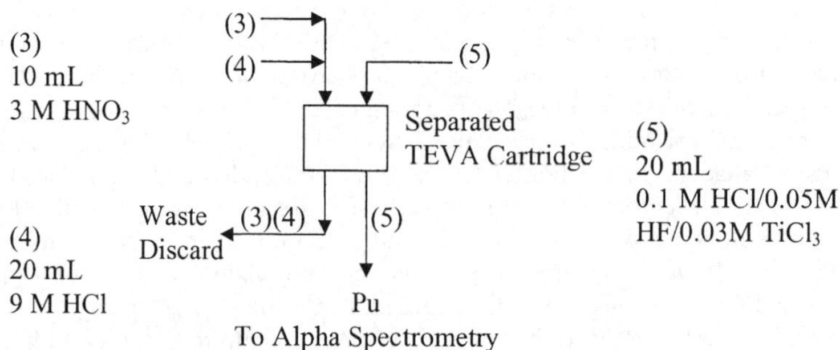

Figure 4. *Flowchart for isolation of radiochemically pure Pu fraction*

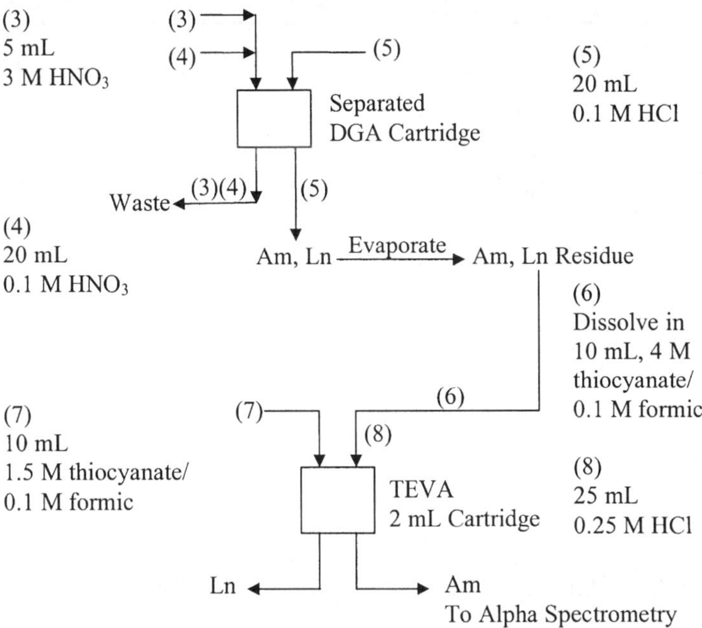

Figure 5. *Flowchart for isolation of radiochemically pure Am fraction*

The overall separation scheme for Pu and Am from preconcentrate is shown in Figures 3, 4, and 5.

2.3 Distribution of Leached Matrix Constituents

The distribution of leached matrix constituents during the DGA column run was measured by performing the entire separation scheme as depicted in Figure 2 except that no tracers were added to the soil. However, 400 μg of Sm(III) was added to reveal the location of Am(III). Each fraction obtained during the column run was then analyzed by ICP/AES.

3 RESULTS AND DISCUSSION

3.1 Preconcentration

Figure 6 shows a chromatogram obtained by processing the leachate from 100 grams of Northern Illinois topsoil. The soil was spiked with a known quantity of ^{239}Pu and ^{241}Am. One can see that there is negligible loss of Pu and Am during the load and rinse steps. Ten bed volumes (20 mL) of stripping gives a 75% yield based on the quantity of Pu and Am spiked on the 100 grams of soil. Separate experiments on the same soil where only the leachate itself was spiked with activity showed that >99% of the Pu and Am was recovered from the load solution. We, therefore, conclude that ~25% of the activity spiked onto the soil is not recovered. (However, we will show below in experiments with reference soils

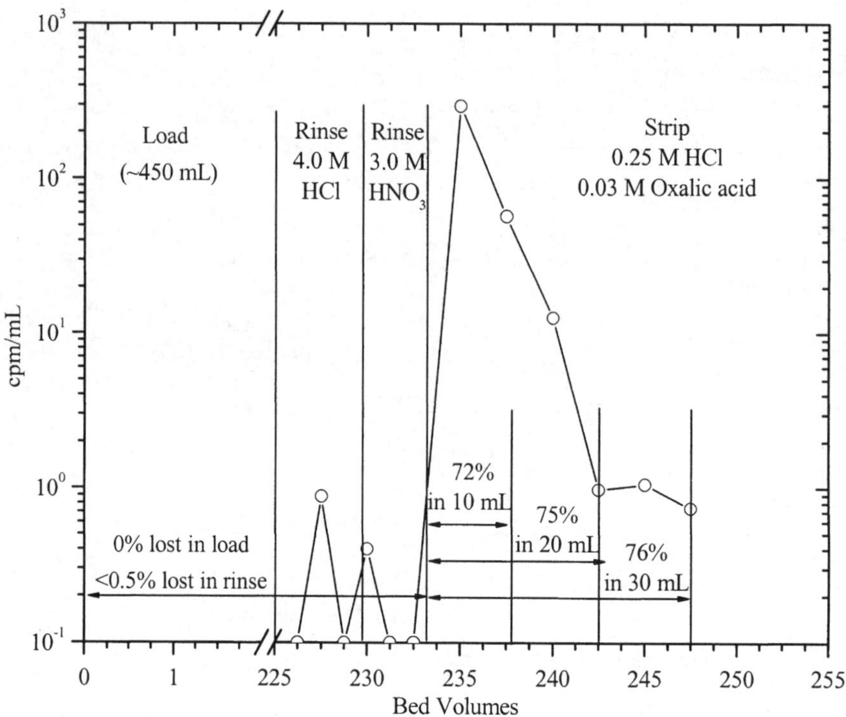

Figure 6. *Elution profiles of ^{239}Pu and ^{241}Am on 2 mL DGA preconcentration column. Column dimensions 1.5 cm i.d. by 1 cm in length. Flow rate is 7 to 8 mL/min.*

that the recovery of Pu and Am already in the soil and the activity that we spike to obtain yields are the same.) The entire column run requires ~60 minutes. Table 1 shows the distribution of leached matrix constituents in the DGA column run. Except for the lanthanide elements and calcium, greater than 99.5% of all matrix constituents report to the load and rinse solutions. The small quantity of Ca found in the strip solution is most likely a result of the small synergistic enhancement of Ca^{2+} uptake in the presence of $FeCl_3$. It is interesting to note that the HNO_3 rinse effectively strips Fe(III) which is consistent with the behaviour of Fe(III) in nitrate media[7]. The combined Ca and Fe found in the HNO_3 rinse and strip solutions most likely represents the capacity of the DGA column assuming a 1:1 stoichiometry. The level of Sm found in the strip solution is >400 μg because of the Sm naturally present in the soil.

3.2 Overall Recoveries

Tables 2, 3, and 4 show the overall recoveries of Pu and Am obtained with spiked 100 g samples of Northern Illinois topsoil and two MAPEP reference soils following the separation schemes outlined in Figures 3, 4 and 5. Figures 7 and 8 show the alpha spectra obtained for one of the reference soils. The overall recoveries are in the 50 to 60% range (with one exception). However, if one assumes that at least 25% is lost in the leaching step, then the overall recoveries in the actual separation steps is a respectable 80 to 90%. In the case of the two reference soils, the agreement between the Bq/Kg found for the ^{238}Pu and

^{241}Am is in the range of 5 to 10% after correction for the recoveries of the ^{242}Pu and ^{243}Am yield tracers.

Table 1 *Mass (mg) of leached matrix constituents in DGA column run fractions*

Element	Load[a]	HCl Rinse[b]	HNO_3 Rinse[c]	Strip[d]
Al	1240	5.9	0.12	0.044
Ba	6.9	0.06	0.001	<MDL
Ca	136	5.5	10	5.4
Cu	2.0	0.01	0.0005	<MDL
Fe	1770	14	19	0.049
Mg	3160	16	3.3	0.055
Mn	3.8	0.2	0.0055	<MDL
Na	19	<MDL	<MDL	<MDL
Ni	2.5	0.006	0.002	<MDL
P	118	0.64	0.014	<MDL
Pb	4.1	0.02	0.007	<MDL
Si	5.3	0.045	0.065	0.06
Ti	15	0.072	<MDL	<MDL
La	0.32	0.25	0.13	0.77
Ce	0.61	0.21	0.054	0.48
Pr	0.45	0.0042	0.0079	0.30
Nd	0.22	0.018	0.0077	1.14
Sm	<MDL	<MDL	<MDL	0.47
Eu	<MDL	<MDL	<MDL	0.016

[a] 450 mL of ~3.5 M HCl

[b] 10 mL of 4 M HCl

[c] 5 mL of 3 M HNO_3

[d] 20 mL of 0.25 M HCl + 0.03 M Oxalic acid

<MDL = less than minimum detectable level

4 CONCLUSION

A rapid method for the preconcentration of non-refractory Am and Pu from 100 gram soil samples has been developed. The key step in the process is the strong retention of Am and Pu by a diglycolamide extraction chromatographic resin from the moderate to high concentrations of hydrochloric acid in the presence of $FeCl_3$. Overall recoveries of Am and Pu excluding leaching are in the range of 80 to 90%. Recoveries including the leaching step are in the 50 to 60% range. The quantities of non-refractory ^{238}Pu and ^{241}Am found in two different reference soils were within 5 to 10% of the concentrations of above isotopes reported for these soils. Since Fe is ubiquitous in soils, there is usually sufficient Fe to realize the synergistic effect. However, if there is doubt based on the color of the leachate, $FeCl_3$ should be added to the level of 1,000 ppm. Higher concentrations of $FeCl_3$

are even more beneficial. The overall processing time is ~5 hours. Two and one half hours are required for leaching, centrifugation and filtration. The preconcentration step requires one hour and all subsequent chemical separations and source preparations require an additional hour.

The magnitude of the synergistic effect (shown in Figure 1) is unusually large. The most likely explanation is that a portion or all of the chloride counter ions in the Am chloro-DGA complex are replaced by the much larger, more hydrophobic $FeCl_4^-$ complex. Further studies are underway to elucidate the nature of the synergy.

Table 2 *Performance of rapid soil method using 100 g of Northern Illinois top soil*

(processing time – 5 hours)

Isotope	% of Spike Recovered
^{242}Pu	65 ± 2
^{243}Am	57 + 3

Table 3 *Performance of rapid soil method using 100 g of MAPEP-04-MaS12*

(processing time – 5 hours)

Isotope	Found* (Bq/kg)	MAPEP Value (Bq/kg)
^{238}Pu	31.2	35.4
^{241}Am	68.4	67.0

* Based on 35.5% ± 1 recovery of ^{242}Pu and 61.5% ± 1 recovery of ^{243}Am spikes

Table 4 *Performance of rapid soil method using 100 g of MAPEP-05-MaS14*

(processing time – 5 hours)

Isotope	Found* (Bq/kg)	MAPEP Value (Bq/kg)
^{238}Pu	57.0	60.8
^{241}Am	107	94.0

* Based on 63% ± 2 recovery of ^{242}Pu and 54 ± 1 recovery of ^{243}Am spikes

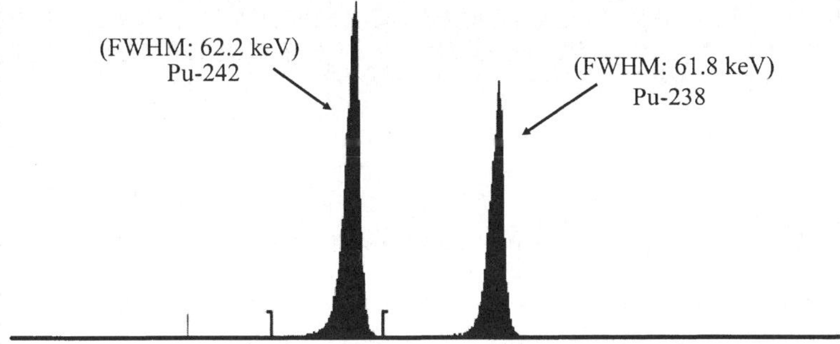

Figure 7. *Alpha spectrum of Pu fraction from 100 gram soil.*

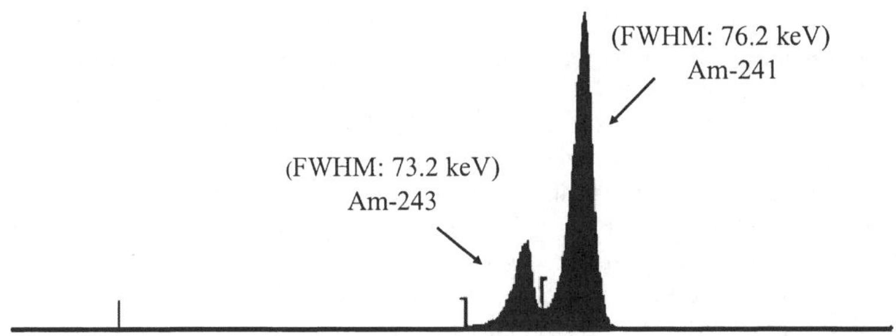

Figure 8. *Alpha spectrum of Am fraction from 100 gram soil.*

References

1 L. L. Smith, J. S. Crain, J. S. Yaeger, E. P. Horwitz, H. Diamond and R. Chiarizia, *J. Radioanal. Nucl. Chem.* 1995, **194**, 151
2 W. C. Burnett, D. R. Corbett, M. Schultz, E. P. Horwitz, R. Chiarizia, M. Dietz, A. Thakkar and M. Fern, *J. Radioanal. Nucl. Chem.* 1997, **226**, 121
3 G. Kim, W. C. Burnett and E. P. Horwitz, *Anal. Chem.* 2000, **72**, 4882
4 H. Dulaiova, G. Kim, W. C. Burnett and E. P. Horwitz, *Radioact. Radiochem.* 2001, **12**, 4
5 E. P. Horwitz, M. L. Dietz, R Chiarizia, H. Diamond, S. L. Maxwell, III and M. R. Nelson, *Anal. Chim. Acta* 1995, **310**, 63
6 S. L. Maxwell, III, *J. Radioanal. Nucl. Chem.* 2006, **267**, 537
7 E. P. Horwitz, D. R. McAlister, A. H. Bond and R. E. Barrans, Jr., *Solvent Extr. Ion Exch.* 2005, **23**, 319
8 C. W. Sill, *Nucl. Chem. Waste Manage.* 1987, **7**, 201

IMPROVEMENTS IN UNDERGROUND GAMMA-RAY SPECTROMETRY AND THE APPLICATION OF MEASURING RADIOACTIVITY IN AGRICULTURAL SAMPLES

P. Lindahl, M. Hult, F. Cordeiro, J. Gasparro, A. Maquet, G. Marissens and P. Kockerols

European Commission, Joint Research Centre, Institute for Reference Materials and Measurements (EC-JRC-IRMM), Retieseweg 111, B-2440 Geel, Belgium

1 INTRODUCTION

High Purity Germanium detectors (HPGe-detectors), for performing gamma-ray spectrometry, could be said to be today's workhorses for measuring environmental samples in any radionuclide laboratory. One reason for this is that the sample preparation for gamma-ray spectrometry is often easy and not very labour intensive. It is also a technique that is capable of delivering activity results in Bq for many radionuclides in one single measurement. The decrease of anthropogenic radioactivity in the environment has highlighted the need for new sensitive measurement techniques. Improvements in germanium productions have resulted in bigger HPGe-detectors with higher efficiency. To further reduce detection limits it is essential to have a low background. The low background that can be obtained by performing gamma-ray spectrometry underground is useful in a wide range of applications such as measurements of double beta decay, safeguards work, radiopurity measurements and environmental radioactivity.[1,2]

IRMM (Institute for Reference Materials and Measurements) is performing gamma-ray spectrometry in the underground laboratory HADES and is therefore continuously involved in work on improving the background of germanium detectors. This paper presents background improvements of an underground HPGe-detector. This detector and a detector located above ground at IRMM were used in a project for the measurement of radioactivity in wheat samples with the aim of distinguishing differences in isotopic fingerprints from organic and conventional farming. Isotopic fingerprinting using multi-dimensional data has recently been used in similar studies involving e.g. determination of the origin of nuclear fuel[3], stable elements in wine[4] and barley[5].

Organic food and farming is an expanding market for food producers in Europe due to the growing consumer interest for certified organic products. For this reason, various laboratories have started research programmes on different analytical techniques for authenticating organic food products. The main idea behind measuring radionuclides in farming products for authenticity purposes is the difference in use of fertilizers for the two farming systems. The uptake of radionuclides in plants is known to be affected by physical, chemical and biological conditions of the soil as well as the individual chemical properties of the nuclide.

2 MATERIALS AND METHODS

2.1 HPGe-detector

The underground HPGe-detector presented here is installed in the underground research facility HADES. It is located in a clay layer at a depth 225 m below ground-level, which corresponds to 500 m w.e. (water equivalent), at the Belgian Nuclear Research Centre (SCK•CEN) site in Mol, Belgium, and is operated by EURIDICE. The detector is an ultra low background detector constructed using materials selected for a high degree of radiopurity as well as for underground usage.

It is a coaxial detector with a relative efficiency of 105% and was produced by CANBERRA Semiconductor n.v. in 2000. It is a so called extended range detector with a p-type crystal and a submicron deadlayer (0.5 μm). It has also a high purity aluminium endcap. The entrance window thickness is 1.5 mm. The active volume of the crystal is 400 cm^3 and the crystal diameter is 80 mm. The detector was first installed with a copper endcap with a carbon epoxy window glued to it. This paper describes the background reduction that was achieved by changing the endcap.

Figure 1 shows a schematic drawing of the underground HPGe-detector and its shield and with a Marinelli beaker sitting on the detector endcap. The inner part of the shield is made from electrolytic copper and is about 10 cm thick. Such a thick copper lining is clearly inappropriate for use above ground because of the cosmic ray interactions, but at 500 m w.e. it is suitable. The inner 5 cm of lead has a ^{210}Pb massic activity of 2.5 Bq/kg and the outer 10 cm has a ^{210}Pb massic activity of 20 Bq/kg. The shield is tight in order to minimise the influx of radon in air and it is flushed with N_2 that boils off from the LN_2 dewar.

2.2 Determination of background location

In ultra low level gamma-ray spectrometry it is not enough to know the background count rate of peaks. It is also essential to know the location of the source of the background, at least whether it is inside the detector or outside (shield or environment). The reason for this is that a massive sample in e.g. a Marinelli beaker can significantly reduce the background count rate from sources outside the detector. There can also be a reduction in background peak count rate for certain peaks from radionuclides with cascading gamma-rays like ^{60}Co located inside the detector. The reason for this is so called summing out, in which a photon that is Compton scattered from the sample back into the detector results in a pulse that is summed with the full energy peak (FEP) of the other cascading gamma-ray. One way of knowing the background sources is to measure the radiopurity of all the components that go into the detector. If this is not possible one can try to shield the detector with a radiopure substance and record the difference in count rate. Kaye et al. (1973) used pure mercury as a shielding substance for a Ge(Li) detector above ground with a resulting overall background reduction of about 18%.[6] In this work a 0.5 L Marinelli beaker (height: 10 cm, outer diameter: 12 cm, cavity diameter: 10 cm and cavity depth: 8.5 cm) was filled with 7.5 kg of hyperpure 6 N mercury (99.9999%). The mercury thickness on top of the detector was about 3 cm and on the side of the detector about 1 cm. Mercury has a high linear attenuation coefficient and the transmission of e.g. 100 keV, 500 keV, 1500 keV and 2000 keV gamma-rays through 1 cm of mercury is 0.01%, 12%, 50% and 53%, respectively. The mercury filled Marinelli beaker was placed around detector Ge-4 and measured for 88 days. A third way of obtaining information on background locations is to measure the ratio between the sum peak and the full energy peaks for gamma-rays from cascading radionuclides. The drawback with this technique is that it requires very long collection times as the background count rate of sum peaks is generally very low. In the Heidelberg-Moscow double beta decay experiment this technique was used to identify certain background sources, but 15 years of data were used for this.[7]

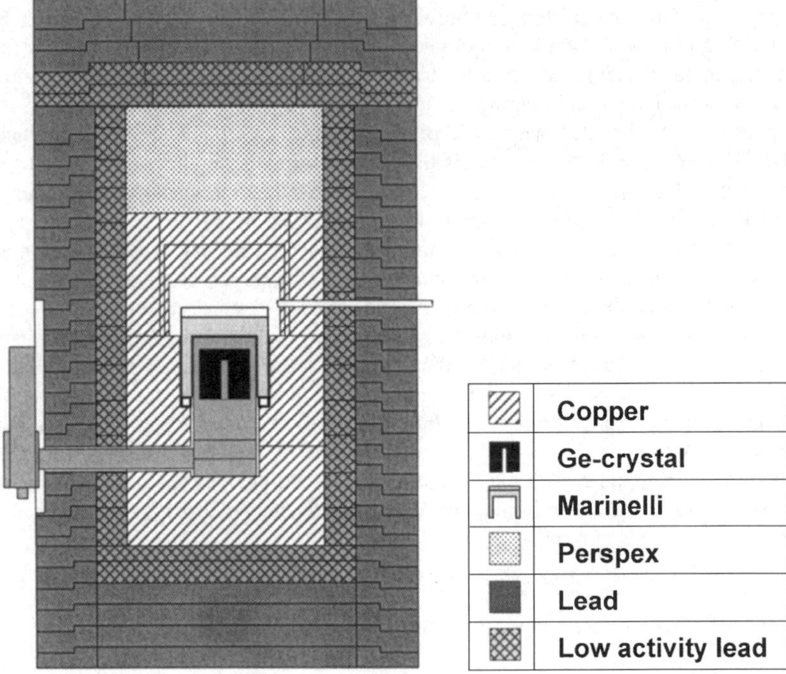

Figure 1 *Schematic drawing of the underground HPGe-detector and the Pb/Cu shield together with Marinelli beaker.*

2.3 Plant Materials

Two agricultural systems (organic and conventional) containing winter wheat (*Triticum aestivum L.*) were investigated in this study. They were grown on two different fields that were in the same region (~8.2 km apart). The organic field was located in Barvaux-en-Condroz, Belgium and the conventional field was located in Monin (Achet), Belgium. The conventional field was fertilized with a liquid N fertilizer and in the organic field an ordinary manure fertilizer was used. Within three parcels (~8 x 8 m) from each field, 35-40 plants were collected during the summer of 2005. Six samples (3 organic and 3 conventional) of grain, stem and the root were prepared and measured. After cutting the plants, the different parts of the plants were dried for 24 hours at 50 °C. Thereafter they were homogenised, weighed and placed in airtight Teflon measurement containers.

For each field, four soil subsamples (layer 0-30 cm) were extracted using a drill and mixed into one sample of about 1 kg. The soil samples were also dried for 24 hours at 50 °C and homogenised, weighed and placed in Teflon measurement containers.

3 RESULTS AND DISCUSSION

3.1 Background Improvement

In a previous study[8] it was found that the carbon epoxy material in the entrance window of the detector contained about 3.2 Bq kg^{-1} of ^{40}K and it was concluded that it should be worth while to substitute the window or the endcap. It was assumed that the copper endcap was not a major source of background but in order to maintain the high efficiency for low energy gamma-rays it was not possible to use a copper window. High purity aluminium (6N) has proven to be a radiopure material with little attenuation of gamma-rays. It was, however, not possible to simply replace the carbon epoxy window with an aluminium window as this required gluing and a suitable radiopure glue is not possible to find.[9] It was decided to improve the background by changing the complete endcap into an aluminium one.

The background spectra (0-1400 keV) before and after the replacement of the endcap are shown in Figure 2 and an analysis of the count rates is given in Table 1. The replacement resulted in an improvement of the ^{40}K background by a factor of 3.5 and for the ^{226}Ra background it was a factor of 2. A slight, but not significant, increase in the peaks associated to ^{232}Th was observed. This may be explained by the fact that aluminium is known for containing ultra trace impurities of Th. Furthermore, the overall background count rate below the 1460 keV peak (40-1400 keV) decreased by about 40%. The two peaks 811 and 818 keV in the 2005 spectrum are due to traceable amounts of ^{58}Co in the crystal and the cryostat, which was formed by neutron activation of Cu as well as spallation reactions in the Ge-crystal when the detector was above ground for the assembling of the new endcap.

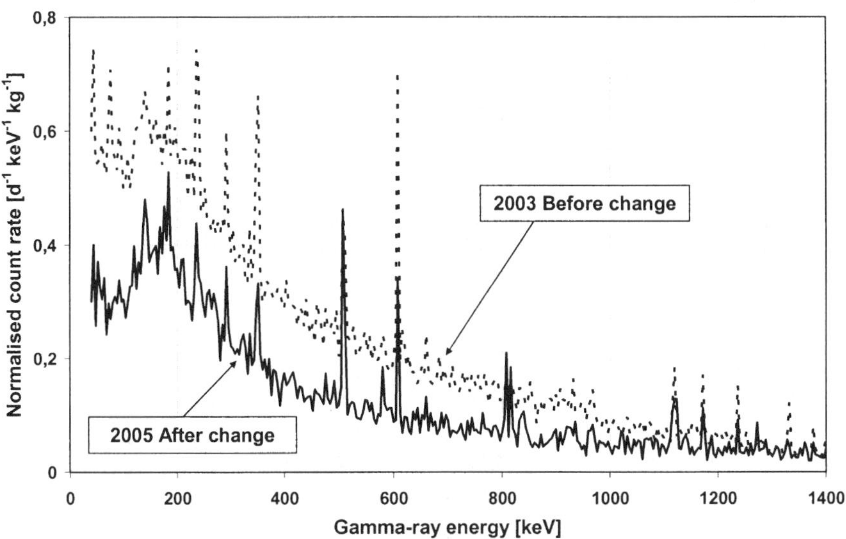

Figure 2 *Background spectra of the underground HPGe-detector before and after the change to an aluminium endcap.*

The mercury shield in the Marinelli beaker did not cover a complete 4π solid angle of the crystal but it should anyhow give an indication of the location of background components and it was the most suitable way of determining the self shielding effect of a sample in a similar Marinelli beaker. Table 1 shows that the count rates with the mercury shield were unchanged for peaks associated with ^{210}Pb and ^{60}Co indicating that the main location for these radionuclides is inside the detector. A significant reduction was observed for the 186 keV peak, where the larger part is expected to come from ^{226}Ra. The ^{222}Rn-daughters, ^{214}Pb and ^{214}Bi, also showed a significant reduction indicating the major source of ^{226}Ra is outside the detector. The background count rates in the peaks from radionuclides belonging to the ^{232}Th decay series (^{208}Tl, ^{212}Pb and ^{228}Ac) and ^{40}K was increasing with the mercury shielding. This contribution is coming from tape sitting outside the mercury container. The tape may also be a source of ^{226}Ra contribution to the peaks of ^{222}Rn-daughters.

Table 1 *List of background γ-ray peaks for the underground HPGe-detector before (2003) and after (2005) the installation of the new endcap as well as with mercury shield (2006-Hg). The combined standard uncertainty is given within brackets.*[10] *Decision thresholds are given according to ISO.*[11] *Reference date: June 1, 2005*

Eγ (keV)	*Radionuclide*	*2003* (counts d^{-1})	*2005* (counts d^{-1})	*2006-Hg* (counts d^{-1})
46	^{210}Pb	1.2 (4)	1.19 (26)	1.1 (3)
63	^{234}Th	< 0.8	< 0.8	< 0.5
93	^{234}Th	< 0.9	< 0.7	< 0.6
186	^{226}Ra+^{235}U	1.9 (6)	1.0 (4)	< 0.6
238	^{212}Pb	2.5 (4)	1.02 (28)	1.38 (21)
242	^{214}Pb	1.2 (4)	0.38 (19)	0.41 (16)
295	^{214}Pb	1.9 (4)	1.04 (26)	0.6 (3)
338	^{228}Ac	< 0.8	0.4 (3)	0.48 (29)
351	^{214}Pb	4.5 (4)	2.05 (28)	1.72 (18)
583	^{208}Tl	0.92 (27)	0.63 (17)	1.10 (19)
609	^{214}Bi	4.4 (4)	2.38 (27)	1.53 (20)
661	^{137}Cs	0.58 (22)	0.29 (17)	< 0.3
911	^{228}Ac	0.29 (21)	0.38 (18)	1.34 (17)
969	^{228}Ac	< 0.5	0.39 (14)	0.61 (12)
1120	^{214}Bi	1.41 (20)	0.85 (17)	0.53 (11)
1173	^{60}Co	0.66 (12)	0.59 (14)	0.54 (14)
1332	^{60}Co	0.69 (11)	0.34 (13)	0.39 (13)
1460	^{40}K	10.5 (4)	3.09 (28)	4.07 (23)
1764	^{214}Bi	1.23 (15)	0.70 (14)	0.77 (13)
2614	^{208}Tl	0.78 (12)	0.86 (15)	1.76 (15)
	40-2700 keV	766 (3)	456 (3)	468.0 (23)
	40-1400 keV	708 (3)	409 (3)	412.9 (22)
	1500-2700 keV	39.9 (8)	37.4 (9)	43.2 (7)
	Measuring time	66 days	45 days	88 days

The source of the ^{60}Co signal could be partly from the Ge-crystal as well as from the Cu holder and cryostat as both the copper as well as the germanium were stored above ground and activated by cosmic rays before the detector was assembled and brought underground. The copper in the shield was only a very limited time above ground (about 2 weeks) during transport from the producer and during machining, which would explain the low contribution to the ^{60}Co peaks from outside the detector. The overall integrated background count rate increased slightly when using the mercury shielding, mainly due to the contribution from the tape.

3.2 Radionuclides in organic and conventional wheat samples

Figure 3 shows the spectra from measurement of a wheat stem sample with two different detectors, above ground and underground. It clearly shows the main advantage of using underground detectors for low level environmental samples. The peaks above the 1460 keV peak can be measured with very low detection limits.

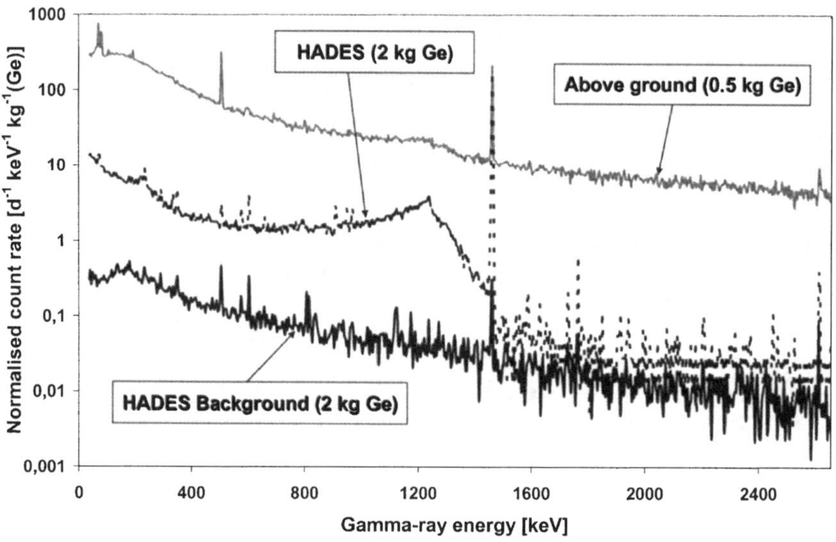

Figure 3 *Gamma ray spectra from the same wheat sample (stem) measured in HADES and above ground at IRMM. The background spectrum from HADES is shown as comparison.*

The grain samples were only measured above ground and thus only ^{40}K data are available at this time. There was no significant difference observed between the two agricultural systems with respect to the ^{40}K content in grain samples. The average massic activity for the organic and conventional field was 128.9(21) Bq kg^{-1} and 133(9) Bq kg^{-1}, respectively. The N fertilizer treatment appears not to influence the uptake of ^{40}K in the wheat grains with an organic to conventional ratio of 0.97(7). This can be compared to Pulhani et al. (2005) who found an organic to conventional ratio of about 1.2.[12] The organic field in that study was treated with ammonium phosphate fertilizer.

Tables 2-4 show the results of the gamma-ray measurements of soil, wheat root and wheat stem from the two agricultural systems.

The ^{40}K content in soil shows a significant difference between the two systems with lower massic activity for the organic field. The opposite relation was observed for ^{137}Cs. For the radionuclides from the natural decay series a slight trend of lower massic activity in the organic field was observed. For the root samples a significant difference between the two fields was observed for all the radionuclides with higher activities from the organic field. Earlier study has shown that ammonium and potassium ions can cause significant reductions in the uptake of Cs in wheat roots.[13] The results of the stem samples were different compared to the root samples with higher concentrations for the conventional field. The explanation for this is still yet to be investigated.

Table 2 *Average massic activities (Bq kg^{-1}, dry weight) of radionuclides in soil samples from conventional and organic agricultural systems for wheat farming, 2005. The combined standard uncertainty is given within brackets.*[10]

Radionuclide	*Conventional* [Bq kg^{-1}]	*Organic* [Bq kg^{-1}]	*Org/Conv ratio*
^{137}Cs	3.20 (17)	5.54 (28)	1.73 (13)
^{40}K	440 (23)	255 (13)	0.58 (4)
^{235}U	1.70 (23)	1.45 (22)	0.85 (17)
^{238}U	28 (4)	28 (4)	0.99 (21)
^{226}Ra	19 (3)	17.9 (24)	0.93 (18)
^{210}Pb	42 (4)	38 (4)	0.89 (13)
^{228}Ra	26.1 (17)	20.3 (8)	0.78 (6)
^{228}Th	25 (6)	19 (4)	0.79 (26)

Table 3 *Average massic activities (Bq kg^{-1}, dry weight) of radionuclides in wheat (root) samples from conventional and organic cultural systems, 2005. The combined standard uncertainty is given within brackets.*[10]

Radionuclide	*Conventional* [Bq kg^{-1}]	*Organic* [Bq kg^{-1}]	*Org/Conv ratio*
^{137}Cs	0.96 (23)	4.68 (14)	4.9 (12)
^{40}K	253 (33)	336 (11)	1.32 (18)
^{235}U	0.32 (6)	1.09 (14)	3.4 (7)
^{238}U	4.6 (6)	12.2 (5)	2.7 (4)
^{226}Ra	6.4 (13)	15.9 (11)	2.5 (5)
^{210}Pb	12.3 (15)	21.0 (10)	1.70 (22)
^{228}Ra	8.4 (5)	19.0 (8)	2.26 (16)
^{228}Th	6.0 (8)	16.26 (26)	2.7 (4)

Table 4 *Average massic activities (Bq kg^{-1}, dry weight) of radionuclides in wheat samples (stem) from conventional and organic agricultural systems, 2005. The combined standard uncertainty is given within brackets.*[10]

Radionuclide	*Conventional* [Bq kg^{-1}]	*Organic* [Bq kg^{-1}]	*Org/Conv ratio*
^{40}K	736 (28)	338 (14)	0.460 (26)
^{226}Ra	5.19 (26)	0.98 (7)	0.188 (16)
^{228}Ra	5.7 (4)	1.02 (15)	0.18 (3)
^{228}Th	2.19 (20)	0.864 (26)	0.39 (4)

In Table 5 the observed transfer factors (TF) for roots, stems and grains are listed for the two agricultural systems. The TF is calculated as the ratio of the massic activity in plant material to its massic activity in soil. Generally a lower TF for root samples from the conventional system was observed indicating lesser uptake from soil to root of the radionuclides in this study.

Table 5 *Observed transfer factors, TF (Bq kg^{-1} plant / Bq kg^{-1} soil, dry weight) in different parts of wheat samples (root, stem and grain) from conventional and organic agricultural systems, 2005. The combined standard uncertainty is given within brackets.*[10]

Element	*Organic*	*Conventional*
Root		
K	1.32 (8)	0.58 (8)
Cs	0.85 (5)	0.30 (7)
Ra	0.91 (3)	0.326 (6)
Th	0.83 (19)	0.25 (6)
Pb	0.55 (6)	0.29 (5)
U	0.59 (22)	0.176 (18)
Stem		
K	1.33 (9)	1.67 (11)
Ra	0.052 (3)	0.24 (4)
Th	0.044 (10)	0.089 (22)
Grain		
K	0.506 (27)	0.303 (25)

4 CONCLUSION

- In underground laboratories the main source of background in HPGe-detectors is usually from radioactivity inside the detector. It is possible to improve the background by careful selection of radiopure materials.
- The background reduction presented in this paper, which is in the order of a factor of 2-3, may sound insignificant but has proven important for several projects where the peak count rate is near the decision threshold. In e.g. the case of measuring ^{60}Co in samples from Hiroshima, the background reduction accomplished here was the difference between a positive or negative detection.
- Early results on isotopic fingerprinting using gamma emitting radionuclides shows that it is a promising complementary technique for distinguishing between organically cultivated wheat and conventionally cultivated wheat. A significant difference in uptake of K, Cs, Ra, Th, Pb and U in root and K, Ra and Th in stem was observed between the two agricultural systems.

References

1. M. Hult, W. Preusse, J. Gasparro and M. Köhler, *Acta Chim. Slovenica*, 2006, **53**,1.
2. M. Hult, J. Gasparro, P.N. Johnston and M. Köhler, in proceedings from: *1st International Symposium of Environmental Monitoring and Prediction of Long- and Short-Term Dynamics of Pan-Japan Sea Area: Construction of Monitoring Network and Assessment of Human Effects*, Kanazawa, Japan, March, 2003, p. 18.
3. G. Nicolaou, *J. Environ. Radioactivity*, 2006, **86**, 313.
4. P. Kment, M. Mihaljevič, V. Ettler, O. Šebek, L. Strnad and L. Rohlová, *Food Chemistry*, 2005, **91**, 157.
5. S. Husted, B.F. Mikkelsen, J. Jensen and N.E. Nielsen, *Anal. Bioanal. Chem.*, 2004, **378**, 171.
6. J.H. Kaye, F.P. Brauer, J.E. Fager and H.G. Rieck, *Nucl. Inst. Meth.*, 1973, **113**, 5.
7. O. Chkvovets, private communication.
8. M. Hult, J. Gasparro, L. Johansson, P. N. Johnston and R. Vasselli, in *Environmental Radiochemical Analysis II*, Special Publication No. 291, ed. P. Warwick, Royal Society of Chemistry, Cambridge, 2003, p. 375.
9. J. Busto, Y. Gonin, F. Hubert, Ph. Hubert and J.-M. Vuilleumier, *Nucl. Inst. Meth. Phys. Res. A*, 2002, **492**, 35.
10. ISO/IEC/OIML/BIPM, *Guide to the Expression of Uncertainty in Measurement, 1st Corrected Edition*, ISO, Geneva, Switzerland, 1995.
11. ISO, *Determination of the detection limit and decision threshold for ionizing radiation measurements Part 3*, ISO. 2000, ISO 11929-3:2000.
12. V.A. Pulhani, S. Dafauti, A.G. Hegde, R. M. Sharma and U. C. Mishra, *J. Environ. Radioactivity*, 2005, **79**, 331.
13. G. Shaw and J.N.B. Bell, *J. Environ, Radioactivity*, 1991, **13**, 283.

RESPONSES OF U AND Pu TO MICROBIALLY DRIVEN NITRATE REDUCTION IN SEDIMENTS

M. Al-Bokari[1], C. Boothman[2], G. Lear[2], J.R. Lloyd[2], F.R. Livens[1,2]

[1]Centre for Radiochemistry Research, School of Chemistry, The University of Manchester, Oxford Road, Manchester, M13 9PL, UK
[2]Williamson Research Centre for Molecular Environmental Science, School of Earth, Atmospheric and Environmental Sciences, The University of Manchester, Oxford Road, Manchester M13 9PL, UK

The sediments of the NE Irish Sea contain elevated levels of U and Pu, principally arising from historic, authorised low level waste discharges from the Sellafield reprocessing plant. An annual cycle in Pu solubility, attributed to changes in microbial activity, has previously been observed in intertidal sediments of the Esk estuary. In a microcosm experiment using Esk estuary sediments, nitrate reducing conditions developed in 3 days and Pu and U solubility changed by factors of 3 and 2, respectively. Accompanying shifts in the microbial community have been assessed using DNA-based and RNA-based methods, and correlated with the biogeochemical conditions in the microcosms.

1 INTRODUCTION

Since the 1950s, Sellafield has discharged low-level radioactive wastes to the Irish Sea under authorisation.[1, 2] Many components of the waste rapidly become associated with fine suspended particles, which are then redistributed by seasonal and daily water movements.[3-5] A proportion of these particles carry activity into low energy zones, particularly the river estuaries of the NE Irish Sea basin.[6-8] West Cumbria thus represents a unique opportunity to investigate the behaviour of a range of radionuclides in the environment.[9] These studies suggested that both radionuclide transport and changes in actinide solubility could be affected by factors such as geochemical changes and microbial activity.[10-12] Previous studies[5, 7-9, 13] on the migration of Pu and U in intertidal sediment of the Esk Estuary showed that they are largely immobile but also that a very small fraction (~ 10^{-4}) of the total inventory of Pu and U is present in the solution phase. It also appears that solubility fluctuates over the course of the year by a factor of about 4, and that these changes may be related to microbiological activity in the sediment. This work was carried out in order to test this hypothesis.

2 EXPERIMENTAL METHODS

2.1 Experimental Procedure

Approximately 30 kg of sediment was collected from a stable, vegetated salt marsh on the northern shore of the River Esk, which is known to contain elevated levels of transuranic elements.[5] The most active fraction (10-20 cm depth) was used in these experiments. In order to homogenise the sediment without the excessive disruption to the microbial community which would arise from sieving or drying, the sediment was cut into approx 1 cm cubes and these were mixed by hand. The homogeneity of 500 g (wet weight) subsamples was tested using the X^2 test and found to be satisfactory. Immediately after mixing, 500 g (wet weight) of sediment were weighed into sterilised bottles along with 500 mL of CO_3^{2-}/ HCO_3^- buffer solution (pH 7). The solution was made 20 mM in acetate (as electron donor) and nitrate (as electron acceptor). A parallel set of control experiments which contained neither acetate nor nitrate was also run.

A total of seven bottles was prepared for each batch (three for immediate sampling (T_0), three for sampling after nitrate reduction was complete (T) and one (monitoring) to allow monitoring of geochemical indicators). On sampling, the contents of the bottles were centrifuged (5000 rpm, 30 min). The supernatant was retained for actinide analysis, and the sediment for molecular biology analysis. Samples were stored at -80°C where necessary. Radiochemical analysis was carried out for Pu and U using established methods.[14, 15]

2.2 Microbiological Analysis

DNA was extracted from (0.25g) wet sediment using a PowerSoil DNA Isolation Kit (MO BIO Laboratories Inc., Solana Beach, CA).[16] A fragment of the 16S rRNA gene, approximately 520 b.p., was amplified from samples using the broad-specificity primers 8F and 519r and PCR technique using a BioRad iCycler (BioRad, Hemel Hempstead, Herts, UK)[16]. Restriction fragment length polymorphism (RFLP) analysis was performed with restriction endonucleases Sau3A and Msp I and, on the basis of the RFLP results, DNA was selected and sequenced by the dideoxynucleotide method.[16] The feedback sequences were analysed against the NCBI (USA) database using the BLAST program package and matched to known 16S rRNA gene sequences.[16, 17] RNA was selectively extracted from sediment using a modified method[18] excluding cetyltrimethylammonium bromide (CTAB) from the buffer solution. Extracted nucleic acids were resuspended in sterile, nuclease-free water and treated with DNase (RQ1 RNase-Free DNase, Promega, WI., USA) to remove genomic DNA contamination. cDNA was synthesized using the treated RNA samples and the necessary reverse primers with an avian myeloblastosis virus (AMV) reverse transcriptase (Promega) according to the manufacturer's instructions. PCR reactions, RFLP and sequencing were carried out for 16S rRNA as before.

3 RESULTS AND DISCUSSION

3.1 Monitoring Redox State

It is generally[19] accepted that biological systems will exploit terminal electron acceptors (TEA) in order of decreasing Free Energy yield. The amount of energy associated with acetate oxidation using nitrate according to the following biochemical reaction is -806 kJ mol^{-1} CH_3COO^-:

$$CH_3COO^- + 1.6NO_3^- \rightarrow 2CO_2 + 0.8N_2 + 2.6OH^- + 0.2\ H_2O$$

As a result, nitrate is the first TEA to be consumed after the onset of anaerobic conditions and, in these microcosms, nitrate reduction was completed (NO_3^- concentration decreased from 20 mM to 0.04 mM) in 3 days.

3.2 Radionuclide Behaviour

The radionuclide content of the aqueous phase will be a more sensitive indicator of change than the sediment since only a small proportion of the radionuclide inventory is present in solution.[7, 8, 13, 20-22] Over a three day period, the control batch (Figure 1) showed a 20% increase in dissolved U concentration while the Pu concentration was unchanged. This most probably reflects enhanced UO_2^{2+} solubility in the presence of the carbonate buffer solution. In the nitrate-reducing microcosms, uranium and plutonium activity concentrations increased by a 20% and 75%, respectively (Figure 2). The *t-Test* [23] was used to define the significance of the observed concentration changes (Table 2). The results obtained exceed the critical value of t for 4 degrees of freedom at 95% confidence (p= 0.05), which is 2.78, so the concentration changes are statistically significant.

Uranium solubility is increased even more in the nitrate microcosms, and one possible explanation is the conversion of acetate to CO_2, coupled with nitrate reduction, which would give higher dissolved carbonate concentrations in the nitrate microcosms. However, PHREEQE modelling showed that the higher CO_3^{2-} concentration would not greatly affect uranium speciation in solution and is therefore unlikely to account for the enhanced solubility. Alternatively, as nitrate is reduced to ammonium (NH_4^+), which promotes cation exchange, this could lead to displacement of UO_2^{2+} from surface complexes, which are the predominant uranyl species on mineral surfaces.[24]

Plutonium, which is likely to be present as Pu(IV), will be less susceptible to ion exchange due to its tendency to hydrolyse, but some increase in solubility as a result of ion exchange with NH_4^+ is still quite possible. Redox transformation of Pu is unlikely since the standard redox potential for Pu(IV)/Pu(III) is -0.27 V compared with +0.775 V for NO_3^-/NO_2^-. Moreover, Pu(IV) will not be susceptible to complexation by CO_3^{2-} at the prevailing CO_3^{2-} concentrations.

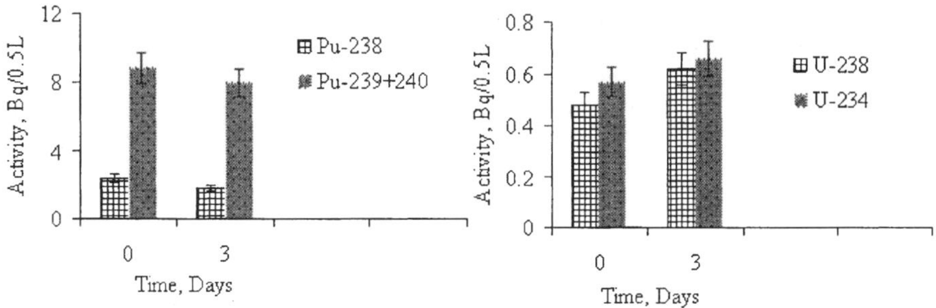

Figure 1: *Changes in Dissolved Pu and U Concentrations in the Control Microcosms*

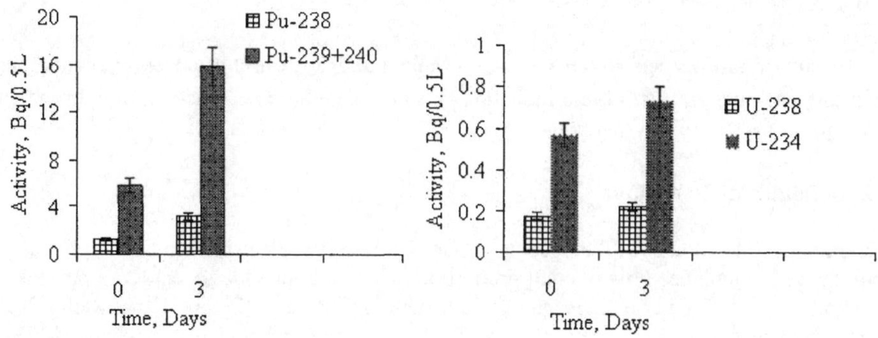

Figure 2: *Changes in Dissolved Pu and U Concentrations in the Nitrate Microcosms*

Table 2: *t-Test values for the Pu and U isotopes*

Batch ID	*t-Test*			
	$^{239+240}$Pu	^{238}Pu	^{238}U	^{234}U
Control	1.12	1.75	15.35	10.11
Nitrate	3.61	4.63	13.86	11.58

3.3 Characterising the Microbial Community

To study the evolution of the microcosms in more detail and follow changes in the microbial community, DNA and RNA profiles for the microbial community at both T=0 and T=3 days were obtained. DNA analysis was carried out to identify those organisms which have existed in the sediment, while RNA reflects the active members of the community. Generally, DNA and RNA profiles gave a wide range of diversity, with over 25 different microorganisms identified.

Table 3 lists the predominant microorganisms in both DNA and RNA profiles for the control batch at T=0. Both DNA and RNA analysis of the nitrate microcosm at T=3 days showed substantial shifts in community composition (Table 4), with *Arcobacter sp.* dominating in both cases. *Arcobacter sp.* are known as nitrate reducers since they play a role in nitrogen fixation, so this shift is consistent with sediment conditions. A couple of additional nitrate-reducing bacteria, *Pseudomonas sp.* and *Nisaea denitrificans,* were also identified.

Table 3: *Dominant Microorganisms in Control Microcosms (T=0)*

Closest relative DNA (Matching Similarity)	Microorganism Percentage in the Community	Number of Base
Dehalococcoides sp. BHI80-1 (90)	14	*(229/264)*
Aeromonas sp. WC56 small subunit (100)	12	*(22/22)*
Defluvicoccus vanus (92)	10	*(368/400)*
Bosea thiooxidans isolate TJ1 (94)	10	*(323/342)*
Endosymbiont of Tevnia jerichonana (96)	8	*(217/225)*
Closest relative RNA (Matching Similarity)	**Microorganism Percentage in the Community**	**Number of Bases**
Benzene mineralizing consortium (86)	8	*(215/249)*
Cytophaga sp. Dex80 (94)	8	*(215/228)*
Methylocystis sp. KS3 (91)	6	*(357/389)*
Geopsychrobacter electrodiphilus (90)	6	*(395/420)*
Thiobacillus prosperus (91)	6	*(263/287)*
Alpha proteobacterium PI_GH2.1.D5 (95)	6	*(424/443)*
Rhodomicrobium vannielii (92)	6	*(409/444)*

Table 4: *Dominant Microorganisms in Nitrate Microcosms (T=3 days)*

Closest relative DNA (Matching Similarity)	Microorganism Percentage in the Community	Number of Bases
Arcobacter sp. KT0913 (94)	62	*(453/480)*
Pseudomonas sp. 2N1-1 (98)	8	*(508/512)*
Geobacter sulfurreducens PCA	*4*	*(185/206)*
planctomycete str. 292 (88)	10	*(407/445)*
Shewanella sp. LT17 (96)	*4*	*(172/178)*
Closest relative RNA (Matching Similarity)	**Microorganism Percentage in the Community**	**Number of Bases**
Arcobacter sp. R-28314 (97)	62	*(477/488)*
Nisaea denitrificans (90)	*12*	*(425/472)*
Pseudomonas sp. 2N1-1 (99)	6	*(510/512)*
Thiobacillus prosperus (91)	4	*(410/443)*
Dehalococcoides sp. BHI80-15 (86)	4	*(228/264)*

4 CONCLUSIONS

In poised microcosm experiments, nitrate reduction was associated with increases in solubility of both U and Pu. In both cases, this probably reflects ion exchange with biogenic ammonium ions. Some increase in U solubility was also observed in control experiments, but this was smaller than in the nitrate-reducing systems and reflects the formation of uranyl-carbonato complexes. Both DNA and RNA analysis showed clear shifts in the microbial community, which was quite diverse before reduction but, after reduction, was dominated by nitrate-reducing bacteria including *Arcobacter sp.*, *Nisaea denitrificans* and *Pseudomonas sp.* Thus, microbiological and geochemical changes can be linked to changes in actinide geochemistry.

References

1 A. S. Hursthouse, M. S. Baxter, F. R. Livens, and H. J. Duncan, *Journal of Environmental Radioactivity*, 1991, **14**, 147.

2 P. J. Kershaw, D. S. Woodhead, S. J. Malcolm, D. J. Allington, and M. B. Lovett, *Journal Of Environmental Radioactivity*, 1990, **12**, 201.

3 J. A. Hetherington, *Marine Science Communications*, 1978, **4**, 239.

4 F. R. Livens and M. S. Baxter, *Journal of Environmental Radioactivity*, 1988, **7**, 75.

5 K. Morris, J. C. Butterworth, and F. R. Livens, *Estuarine Coastal and Shelf Science*, 2000, **51**, 613.

6 F. R. Livens and M. S. Baxter, *Science of the Total Environment*, 1988, **70**, 1.

7 M. J. Keith-Roach, N. D. Bryan, R. D. Bardgett, and F. R. Livens, *Biogeochemistry*, 2002, **60**, 77.

8 M. J. Keith-Roach, J. P. Day, L. K. Fifield, N. D. Bryan, and F. R. Livens, *Environmental Science & Technology*, 2000, **34**, 4273.

9 K. Morris and F. R. Livens, *Radiochimica Acta*, 1996, **74**, 195.

10 Z. Filip and J. J. Alberts, *Science of the Total Environment*, 1994, **144**, 121.

11 C. E. Barnes and J. K. Cochran, *Geochimica Et Cosmochimica Acta*, 1993, **57**, 555.

12 P. E. Kepkay, *Journal of Environmental Radioactivity*, 1986, **3**, 85.

13 F. R. Livens, A. D. Horrill, and D. L. Singleton, *Applied Radiation and Isotopes*, 1992, **43**, 361.

14 N. A. Talvitie, *American Industrial Hygiene Association Journal*, 1971, **32**, 31.

15 N. A. Talvitie, *Analytical Chemistry*, 1972, **44**, 280.

16 A. G. Gault, D. A. Polya, J. M. Charnock, F. S. Islam, J. R. Lloyd, and D. Chatterjee, *Mineralogical Magazine*, 2003, **67**, 1183.

17 I. T. Burke, C. Boothman, J. R. Lloyd, F. R. Livens, J. M. Charnock, J. M. McBeth, R. J. G. Mortimer, and K. Morris, *Environmental Science & Technology*, 2006, **40**, 3529.

18 R. I. Griffiths, A. S. Whiteley, A. G. O'Donnell, and M. J. Bailey, *Applied and Environmental Microbiology*, 2000, **66**, 5488.

19 K. O. Konhauser, R. J. G. Mortimer, K. Morris, and V. Dunn, in 'Role of microorganisms during sediment diagenesis: implications for radionuclide mobility', ed. M. J. Keith-Roach and F. R. Livens, Elsevier, 2002.

20 D. J. Bunker, J. T. Smith, F. R. Livens, and J. Hilton, *Science of the Total Environment*, 2000, **263**, 171.

21 F. R. Livens, A. D. Horrill, and D. L. Singleton, *Estuarine Coastal and Shelf Science*, 1994, **38**, 479.

22 K. Morris, N. D. Bryan, and F. R. Livens, *Journal of Environmental Radioactivity*, 2001, **56**, 259.

23 J. C. Miller and J. N. Miller, 'Statistics for analytical chemistry ', Chichester : Ellis Horwood, 1988, 1988.

24 T. D. Waite, J. A. Davis, T. E. Payne, G. A. Waychunas, and N. Xu, *Geochimica Et Cosmochimica Acta*, 1994, **58**, 5465.

AN EFFICIENT AND OPTIMISED TOTAL COMBUSTION METHOD FOR TOTAL H-3 AND C-14 IN ENVIRONMENTAL AND DECOMMISSIONING SAMPLES

J-S Oh, I Croudace, P Warwick, and D J Kim

GAU-Radioanalytical, National Oceanography Centre, Southampton, SO14 3ZH, UK

1 INTRODUCTION

Tritium needs to be measured during waste characterization of nuclear decommissioning materials since its disposal is subject to the Radioactive Substances Act (1993) Substances of Low Activity (SoLA) Exemption Order criteria. If the total activity is below 0.4Bq/g it is below regulatory concern in the U.K.. During the last five years, the number of total tritium samples analysed by GAU-Radioanalytical increased significantly (Table 1). This is mainly due to the speeding up and expansion of the nuclear decommissioning programme in the UK. One site where GAU-Radioanalytical provided a significant service was at the Winfrith nuclear research site that was home to the prototype heavy-water moderated reactor SGHWR (Lewis et al, 2005). In the UK numerous first and second-generation nuclear facilities are undergoing decommissioning and since April 2005 it has been closely controlled by the Nuclear Decommissioning Authority (NDA). Tritium can be measured using direct aqueous leaching technique and dissolution technique. However, there are limitations using those techniques since leaching is effective for extracting HTO but does not guarantee quantitative extraction of all tritium species and very much depends on the sample types. The dissolution technique involves acids which produce liquid wastes and is labour intensive and time consuming since additional distillation procedures for H-3 and C-14 are normally required. In order to resolve these problems and produce reliable results within short time scale, a technique using a purpose-designed tube furnace was developed by GAU-Radioanalytical. This paper describes the validation of the technique and comparison of several intercomparison exercise results.

2 EXPERIMENTAL

A six tube furnace (Pyrolyser-6 Trio™) supplied by Raddec Ltd. (www.raddec.com) was used for the study (Fig. 1). The furnace consists of three independent heated regions comprising a sample zone, a middle zone and a catalyst zone. The sample zone, where samples are

Table 1. *Number of H-3/C-14 samples*

Year	H-3	C-14
2002	242	79
2003	376	127
2004	847	565
2005	996	153
2006 (by June)	1003	434

normally placed in a silica boat, can ramp up to 900°C. The middle zone switches on automatically and quickly ramps to 500°C when the sample zone reaches its maximum set-point temperature (normally at 500°C but up to 900°C is possible). This rapid heating of the middle zone is designed to remobilize any condensates and transfer them through the catalytic zone where oxidation occurs. The catalyst zone is held at 800°C, and holds a bed of 0.5% Pt on alumina catalyst, and ensures volatile products are quantitatively oxidised to tritiated water and $^{14}CO_2$. The tritiated water is then trapped in a bubbler containing dilute HNO_3 and the CO_2 is trapped in a Carbosorb® bubbler. The ^{3}H and ^{14}C content of the respective bubblers are determined by liquid scintillation counting.

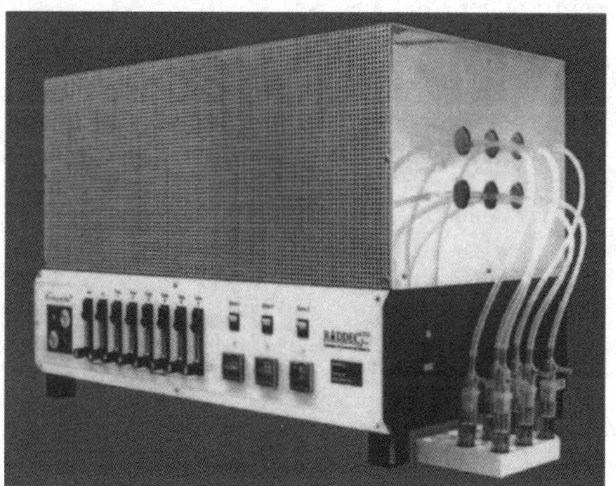

Figure 1. *The Pyrolyser-6 Trio™*

3 H-3 EVOLUTION PROFILES

Some of the H-3 labelled organics together with H-3 in water standard were directly spiked onto a Whatman™ ashless filter paper and combusted in the furnace. Bubblers were changes at regular intervals to establish evolution profiles of those organics. Fig.2 shows evolution profile of those organics and it is evident that all the organic H-3 species are liberated at high temperature (above 300°C) at least over 3 hours. Same tests were performed on both environmental and decommissioning samples previously submitted for analysis. Sample types tested include soil/sediment, fruit, water, grass, milk, fish and sludge as environmental samples. Decommissioning sample types include many materials such as concrete, brick, asbestos, MMMF (fibreglass insulation), metal, plastic, desiccants, paper, electric wire, graphite, paint, wood etc (Fig. 3).

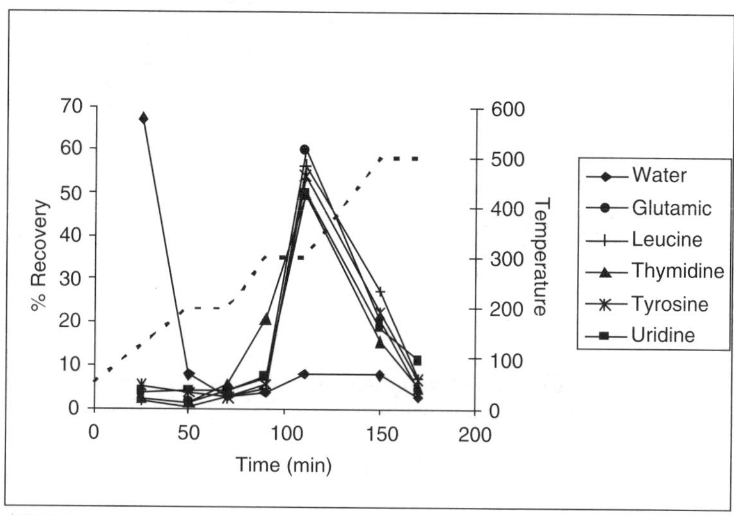

Figure 2. *Evolution profile of H-3 labeled organics*

Apart from the concrete sample, the evolution profiles of other sample types are very similar to those of H-3 labeled organics. This suggests that the major source of contamination for H-3 in concrete sample analysed was tritiated water. The other sample types may have been contaminated by organic H-3 species or H-3 may have bonded within the sample structure resulting in the need of high temperature to liberate H-3. Total run time was determined to be 4 hours as most (>95%) H-3 was released from samples within 4 hours apart from sewage sludge sample. The sewage sludge sample needed minimum of 6 hour combustion. There are two dwelling stages for the protocol, the first at 200°C for 20 minutes and the second at 300°C for 20 minutes. This slow heating is used to prevent the breakage of the silica glass worktube or the end caps, caused by a sudden expansion of air inside the tube when organic rich (or high oil content) samples reach their flash/combustion points. When the sample zone reaches 500°C the air supply is switched over to pure O_2 in order to achieve complete sample oxidation. Catastrophic failures are very rare if the sample heating is carefully controlled.

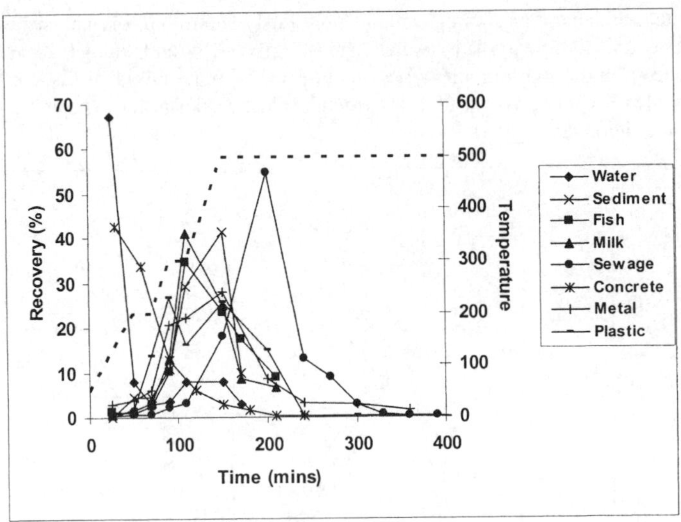

Figure 3. *Evolution profile of H-3 in various sample types. Sewage refers to sewage sludge pellets from the Cardiff East sewage treatment works.*

4 COUNTING EFFICIENCY AND FURNACE RECOVERY FOR H-3/C-14

The furnace was tested for its recovery using aliquots of tritiated thymidine tracer and a C-14 carbonate tracer (Amersham QSA, Harwell). Aliquots of both tracers were added on to a Whatman™ ashless filter paper and placed in one of the furnace work-tube. The optimized 4 hour protocol was used for the recovery test run. Wallac Quantulus™ 1220 Liquid Scintillation Counters, calibrated using certified H-3 and C-14 standards, were used for measurements (Table 2 shows relevant technical information for the Quantulus counters and the resulting yields of the test sample runs).

Table 2. *General technical information*

	SQPE*	Counting efficiency	Furnace recovery**	Limit of detection***
H-3	719 – 729	18 – 20 %	> 90 %	0.020 Bq/g
C-14	720 – 750	66 – 73 %	> 95 %	0.015 Bq/g

* Typical SQPE of concrete samples
** Average value of 70 measurements using an organic H-3 thymidine/C-14 carbonate standard
*** Using 5g sample size and 2 hour counting time on a Quantulus™ (Currie, 1968)

Sample sizes

Using the optimized run protocol, different amount of samples were run to establish the reproducibility. Samples with low proportions of combustible material such as concrete, brick, metal, asbestos etc. show a good H-3 reproducibility for sample masses up to 10 g. However, for samples having high organic contents, such as biota, plastic, organic-rich sediment/soil etc., it was observed that combusting larger sample sizes caused the sudden expansion of the gases in the silica work tube and could displace the end-caps temporarily. As previously mentioned this effect is rare, does not cause breakage and has virtually no impact on the results as the system re-closes almost immediately. Current applications allow combustion of up to 5 g of such samples if the heating is controlled carefully. Since the capacity of Carbosorb-E™ is 4.8 mmol carbon /ml, it is not advisable to combust more than 2 g of biota samples.

Memory effects and cross contamination

A test was carried out to estimate the memory effects after running samples with elevated H-3 activities. Following a sample run and after the sample had been removed, the work tubes were left with a low flow of 50% air:50% oxygen (~0.21 / min). for one hour. A new set of bubblers were then connected to the work tubes and the oxygen was allowed to flow through the system for another 3 hours. The resulting bubbler solutions were counted to determine the memory effect in the system (Fig. 4). This was found to be approximately 3 % and it can normally be removed by leaving the system running overnight with a slow air flow through the work tubes (~0.21 / min). No such memory effects were observed for C-14. There is some evidence that the memory effects are derived from partial tritium retention in the Pt-alumina catalyst bed. Tests showed that no memory effects were observed that originated from bubblers, silicon tubes or sample boats.

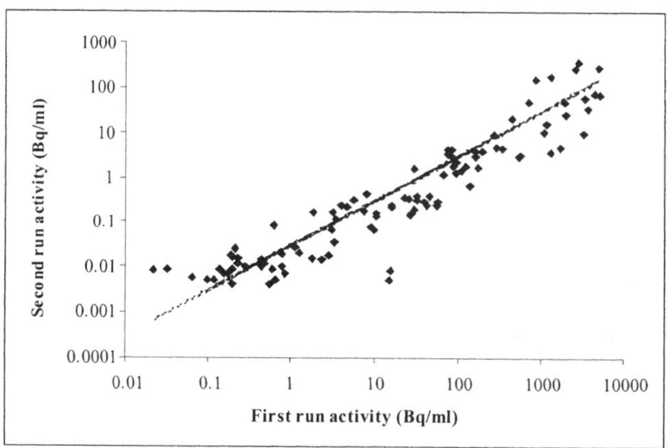

Figure 4. *Memory effect for H-3 (line represents a 3% carry over between runs)*

5 METHOD UNCERTAINTY BUDGET AND ACTIVITY CALCULATIONS

This laboratory reports tritium activities with their total expanded uncertainties (Table 3). The biggest contribution is from the furnace recovery which is 5%. Over 70 standard measurements were used for this. The average value for the furnace recovery is 95±5%. A value of 5.6% expanded uncertainty is used for the final calculation in order to estimate the total expended uncertainty. Standard runs are performed after every five sample runs and the acceptable furnace recovery range is between 80 – 100%. If the furnace recoveries are out of this range, the standard run must be repeated. Blank runs are also performed, especially after the furnace is used for samples with elevated levels of H-3, in order to make sure that there are no cross contamination in the following batch of samples.

Table 3. *Method uncertainty budget*

Parameter	Associated relative uncertainty (1 s.d)
Sample weighing (5g of sample)	0.1%
Decay correction	0.2%
Furnace recovery	5%
Variability in recovery as a result of H-3 speciation	2.5%
LSC calibration curve	1%
H-3 standard (as quoted on certificate)	0.67%
Propagated method uncertainty (1 s.d.)	***5.6 %***

Results for the H-3 Quantulus 1 (8ml + 12ml Gold Star)

Report date : 14-Feb-2006
Customer : GAU
Job reference : STD run
Date samples received : 14-Feb-2006
Date of analysis : 14-Feb-2006
Working instruction number: - GAU/RC/2022
Calibration report number: - GAU/CAL/9006

S/N	Counter ID	Customer S/N	Customer Ref.	Reference date	H-3
1	INST STD	1		14-Feb-2007	616.254 +/- 69.521
2	S-125-1	2		14-Feb-2007	1.519 +/- 0.173
3	S-125-2	3		14-Feb-2007	1.529 +/- 0.174
4	S-125-3	4		14-Feb-2007	1.412 +/- 0.161
5	S-125-4	5		14-Feb-2007	1.522 +/- 0.174

All results are in Bq/g and are decay corrected to the reference date (12.30 Years half life)
"< values" are limits of detection as defined by Currie, 1968
Uncertainties are at the 2 s.d. confidence level and are based on counting statistics alone

Analyst : J.Oh

Figure 5. *Typical print out of LSC-Plus software*

For the final activity calculations a software package called "LSC Plus"(supplied by Raddec Ltd.) is used. This software was developed to avoid using spreadsheet calculations, to minimize transcription errors and to save time. The counting of samples on Quantulus consists of 3 repeat counts. For example, when counting a batch of 20 samples, it produces 60 individual results you then have to manually enter into a spreadsheet. It automatically imports those data and calculates final results and total expanded uncertainties together with the LODs (Fig. 5). It has built-in quality control features introduced to assist in the requirements of ISO17025.

6 INTERCOMPARISON EXERCISE RESULTS

GAU-Radioanalytical has participated in several intercomparison exercises for H-3 and C-14 activities in aqueous and biota samples. These results are in good agreement with the reference values (Table 4). Currently there are no commercially available natural matrix reference materials for total H-3 and therefore the analysis of spiked samples with known certified activities of H-3 is the best possible way of showing that the method is working.

Table 4. *H-3/C-14 intercomparison results*

Supplier	H-3/C-14 type	Measured value (Bq/g ± 1 SD)	Reference value (Bq/g ± 1 SD)
NPL (2002)	Tritiated water	20 ± 1	20.04 ± 0.18
NPL (2004)	Tritiated wated	0.536 ± 0.042	0.539 ± 0.006
NPL (2004)	C-14 carbonate	23.4 ± 1.6	24.4 ± 0.5
FSA	H-3 thymidine (milk)	4.72 ± 0.66	Mean = 4.04 Range 0.18 – 4.93
FSA	H-3 thymidine (plaice)	4.42 ± 0.30	Mean = 4.67 Range 2.7 – 8.3

7 CONCLUSIONS

The PYROLYSER tube furnace has been used successfully to analyse H-3 and C-14 in both environmental and decommissioning samples. The system has been optimised so that most sample types require a four hour run time using an average sample size of five grams. The furnace can combust six samples in one run, and can in principle be used for two runs per day (12 samples). The tritium evolution behaviour has been extensively investigated for a range of typical sample types likely to be encountered during waste characterisation. Such studies are essential to understand the ramping profiles and combustion run times required when using the combustion technique. Typical LODs are 0.020Bq/g and 0.015 Bq/g for H-3 and C-14, respectively. Results from intercomparison exercises show a very favourable agreement with the expected results.

References

L.A. Currie (1968). Limits of qualitative detection and quantitative determination. Anal. Chem., **40** (3), 586-593.

A. Lewis, P.E. Warwick and I.W. Croudace (2005) Penetration of tritium (as tritiated water vapour) into low carbon steel and remediation using abrasive cleaning. J. Radiol. Prot., **25**, 1-8.

THE ANALYTICAL IMPACT ON TRITIUM DATA FROM STORING NUCLEAR DECOMMISSIONING SAMPLES UNDER DIFFERENT CONDITIONS

Dae Ji Kim[1], Ian W. Croudace[1] and Phillip E. Warwick[1]

[1]GAU-Radioanalytical, National Oceanography Centre, University of Southampton, European Way, Southampton SO14 3ZH, UK

1 INTRODUCTION

Tritium is a by-product of civil and military nuclear fission and fusion programmes and radiopharmaceutical production and commonly occurs, though not exclusively, as HTO or organically-bound tritium. During the lifetime of nuclear (involving heavy water) or other active sites tritium compounds may become variably incorporated into the fabric of the buildings. When decommissioning works and environmental assessments are undertaken it is necessary to evaluate tritium activities in a wide range of materials to evaluate waste sentencing options. Compositionally diverse materials (e.g. concrete bioshields, asbestos, wood, desiccants, reactor metal work, graphite blocks and tiles, softwastes etc) will often require characterization and careful consideration needs to be given to their sampling and storage if reliable analytical data are to be achieved. This study represents initial results from a range of simple experiments carried out to investigate tritium emanation rates for several sample types derived from some common reactor waste materials (bioshield concrete, desiccant, and steel). The effects of storing samples at room temperature, in a fridge and in a freezer were considered. All tritium determinations were made using a well-tested combustion method employing a purpose-designed tube furnace followed by LSC.

2 METHOD

Total tritium ($^{3}H_{total}$) was quantitatively extracted from samples using a Raddec Pyrolyser-6 Trio™ System that provides simultaneous oxidation of six samples. The sample in a silica boat was placed in a tube in the centre of the sample zone. The tube end cap was replaced and oxygen-enriched air was passed over the sample. The sample was heated from 50°C to 500°C at a rate of 5°C/minute for 4 hours, with holding stages at 200°C and 300°C for 30 minutes. Pure oxygen was passed through each furnace tube when the sample zone reached 500°C. When required the sample zone was taken to 800 °C, such as during decomposition of graphite. The catalyst zone contained 10 g of Pt-coated alumina catalyst, held at 800°C, which oxidizes any organic compounds in the combustion products to CO_2 and H_2O. The combustion gases are bubbled through a tube containing 20 ml of 0.01 M HNO_3 to trap the combustion water, which contained all of the tritium activity. These methods have been developed and validated by GAU-Radioanalytical at the University of Southampton (www.gau.org.uk).

All tritium measurements were performed using a 1220 liquid scintillation counter (Wallac Quantulus™). 8ml of aqueous sample were mixed with 12ml Gold Star™ (Meridian) scintillation cocktail in a 22ml polythene vial. The counter was routinely calibrated for ^{3}H using a traceable tritiated water standard. In this paper, all uncertainties are quoted at the 90% confidence level except for RO water.

A range of experiments were set up that used active desiccant (310,000Bq/g), concrete (8,950 Bq/g) and metal (110Bq/g) as tritium emanating source materials. RO water, silica gel, cellulose filter paper, plastic, metal, and various desiccants were used as potential receivers. The interaction with drying agents such as silica gel, zeolites and Drierite™ ($CaSO_4$) were also examined.

Figure 1. *Design of the Pyrolyser-6 Trio combustion furnace*

2.1 Experiment 1: Contamination of water by emanation from concrete

Active concrete powder (~ 9 kBq of tritium) was placed in a sealable polythene bag and put in one of eighteen Kilner jars which had very effective rubber lid seals. An open and a closed scintillation vial containing 10 ml each of RO water were added to each of the jars which were then stored in a fridge, a freezer and at room temperature for 1 day, 5 days, 10 days, 15 days, 20 days and 30 days, respectively. At the end of each pre-defined emanation time the respective Kilner jars were opened and the contents removed. 8 ml of each RO water vial (open and closed) was taken and transferred to clean scintillation vials with 12ml of scintillation cocktail (Goldstar) and counted by LSC for 1 hour. Similarly 50 ml of tritium-free RO water was added to the Kilner jar and shaken to collect any tritium (as HTO) that had adsorbed on the walls of the glass jars. An 8ml sample of this was also mixed with scintillant and counted.

2.2 Experiment 2: Tritium emanation from active concrete, desiccant and metal

Samples of active concrete powder, desiccant pellets and metal were placed in sealable polythene bags and then placed in one of nine Kilner jars each having an open and a closed scintillation vial containing 10 ml each of RO water. The jars were each stored in a fridge, a freezer and at room temperature for 2 weeks before being opened after which the scintillation vials were sampled and measured for tritium ingress. Any tritium that had adsorbed on the surface of the Kilner jars was also collected using a 50 ml wash and measured as previously described.

2.3 Experiment 3: Contamination of non-active silica gel, metal, plastic, zeolite and $CaSO_4$

A similar approach to that described above was applied to a range of other materials (silica gel, metal, plastic, zeolite and $CaSO_4$). A known amount of active concrete powder (~ 9 kBq of tritium) was placed in a sealable polythene bag and put in one of fifteen Kilner jars. Each jar contained an open and a closed scintillation vial holding approximately 5g samples of one of the non-active materials. All sample types were prepared in triplicate so that three different storage conditions (freezer, fridge and at room temperature) could be investigated over a 2 week period. At the end of the experiment, each Kilner jar was washed with 50 ml of tritium-free RO water to collect any tritium (as HTO) that had adsorbed on the walls of the glass jars and 8ml of this wash was mixed with scintillant and counted.

2.4 Experiment 4: Tritium emanation from active desiccant

Active desiccant (nominally 170 kBq tritium) was placed in each of twelve sealable plastic boxes having an O-ring lid seal. Samples of (i) silica gel in sealable polythene bag, (ii) silica gel in a double sealed bag, (iii) silica gel in an open scintillation vial and (iv) cellulose filter paper in an open scintillation vial were individually placed in each of the boxes. These were stored for 2 weeks in a fridge, a freezer and at room temperature before opening, sampling and measuring.

3 RESULTS AND DISCUSSION

3.1 Experiment 1: Contamination of water by emanation from concrete powder

The contamination of RO water (open and closed vials) stored with bagged active concrete, under three different storage conditions, was investigated (Figure 2). The greatest contamination transfer occurred to open vials stored at room temperature. Systematically less contamination occurs to open vials stored in a fridge and freezer respectively. Water stored in closed vials at room temperature and in a fridge show significantly less, though measurable, contamination over the 30 day duration of the experiment while the samples in the freezer show no contamination.

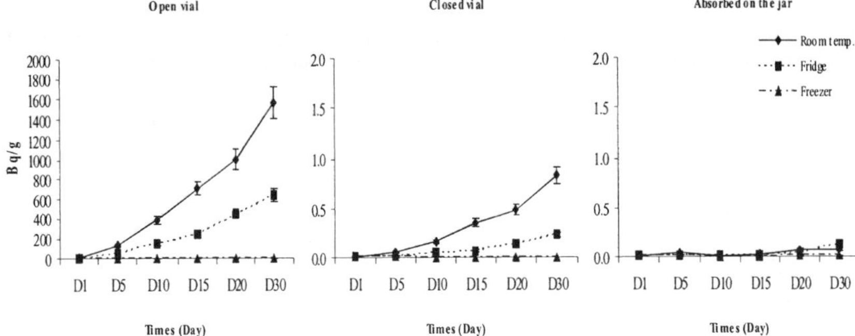

Figure 2. *Variations of cross contamination of tritium from concrete (~9 kBq/g) with the passage of time (Expt. 1). RO water in open and closed scintillation vials were stored in airtight Kilner Jars with a bagged sample of active concrete at room temperature, in a fridge and in a freezer for 1 day, 5 days, 10 days, 15 days, 20 days and 30 days.*

Table 1: *Comparison of cross contamination with source materials and storage temperature (Expt. 1)*

	% contamination transfer to RO water in an open vial			**% contamination transfer to RO water in a closed vial**		
	Rt	**Rf**	**Fr**	**Rt**	**Rf**	**Fr**
Day 1	0.1	<0.1 (0.04)	< 0.007	8.9E-05	8.9E-05	8.9E-05
Day 5	1.5	0.6	< 0.007	5.3E-04	1.0E-04	7.8E-05
Day 10	4.4	1.6	< 0.007	1.8E-03	5.5E-04	5.6E-05
Day 15	8.0	2.7	< 0.007	3.9E-03	7.6E-04	5.6E-05
Day 20	11.2	5.0	< 0.007	5.4E-03	1.6E-03	3.4E-05
Day 30	17.4	7.2	< 0.007	9.2E-03	2.5E-03	5.6E-05

3.2 Experiment 2: Tritium emanation from active concrete, desiccant and metal

Cross contamination of RO water in an open vial following exposure to different materials (bagged) shows a higher contamination than in a closed vial (Table 2). The greatest relative emanation occurs from concrete powder stored but it was also found (checked by triplicate experiments) that the greatest contamination of RO in vials occurred when the Kilner jars were stored in a fridge. This is a different outcome than that found with the concrete from Expt. 1

where RO water stored at room temperature became more contaminated at room temperature than in a freezer and a fridge. All jars containing RO water samples that were stored in a freezer (open and closed scintillation vial) showed very low contamination (Table 2).

Table 2: *Comparison of cross contamination with source materials and storage temperature (Expt. 2)*

Source materials	Storage condition	Activity of source materials (Bq)	Contaminated activity (Bq/g) of RO water					
			RO water open vial	2sd	% contamination transfer to the RO water	RO water closed vial	2sd	% contamination transfer to the RO water
Desiccant	Rt	171120	34.96	3.50	0.2	0.02	0.01	9.9E-05
	Rf	181660	187.19	18.73	1.0	0.09	0.01	4.9E-04
	Fr	157790	0.30	0.03	< 0.1 (0.02)	0.01	-	3.8E-05
Concrete	Rt	24558	66.34	6.64	2.7	0.04	0.01	1.6E-03
	Rf	24882	144.32	14.45	5.8	0.06	0.01	2.2E-03
	Fr	24690	0.70	0.07	< 0.1 (0.03)	0.01	-	2.4E-04
Metal	Rt	263.34	8.61	0.87	32.7	0.01	-	3.0E-02
	Rf	356.62	9.68	0.97	27.1	0.02	0.01	5.0E-02
	Fr	319.33	0.02	0.01	0.1	0.01	-	1.9E-02

Figures in bracket are the relative standard deviation %, Exposure time 2 weeks. Rt: Room temperature, Re: Refrigerator, F: Freezer, - measured values are below the limit of detection; all tests were conducted in triplicate.

3.3 Experiment 3: Contamination of non-active silica gel, metal, plastic, zeolite and $CaSO_4$ by active concrete

Different materials became contaminated to different extents when stored with a tritium emanator such as active concrete (Figure 3). The absorbers used, silica gel, zeolite and Drierite™ ($CaSO_4$), were contaminated to a greater extent than were metal and plastic, especially in an open scintillation vial. Zeolite and Drierite™ ($CaSO_4$) were more contaminated in the fridge (Figure 3). Plastic and metal, however, were relatively little contaminated which implies physical and compositional controls on the extent of contamination [5, 9]. All samples stored in closed vials showed very low contamination.

3.4 Experiment 4: Tritium emanation from active desiccant

Contamination of silica gel and cellulose filter paper stored in a polythene box (having an O-ring lid seal) exposed to emanations from active desiccant were studied (Figure 4). The results show that no significant contamination occurred when containers are stored in a freezer. The same materials stored in a fridge or at room temperature did show increased contaminated with time. These results (Table 2 and Figure 3) show that samples stored in a fridge are more easily contaminated than when stored at room temperature or in a freezer.

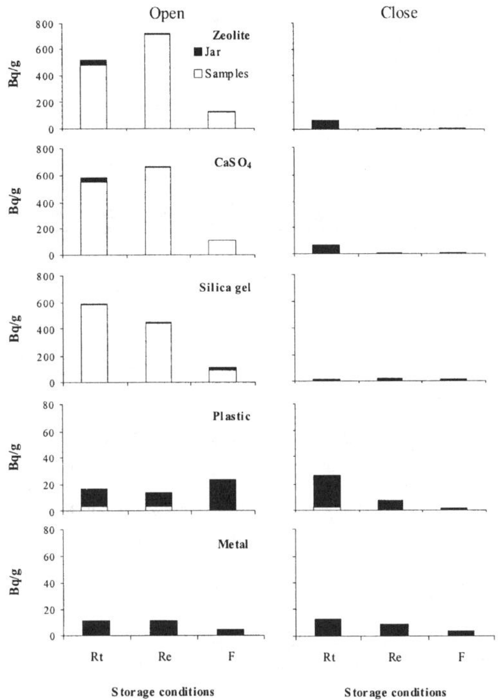

Figure 3. *Comparison of cross contamination of various samples from active concrete (nominally 9 kBq tritium) stored in a fridge, a freezer and room temperature (Expt. 3)*

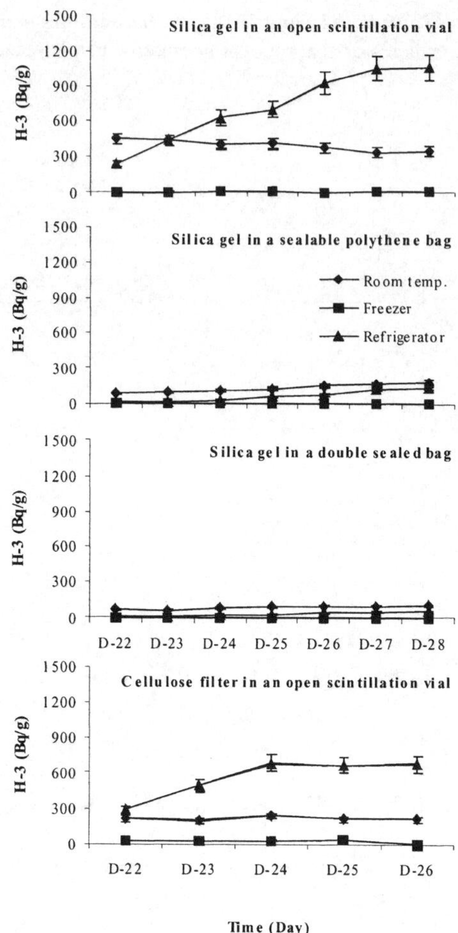

Figure 4. *Variation of cross contamination of silica gel and cellulose filter-paper exposed to an active desiccant emanator (nominally 158 kBq tritium) with time. Storage occurred in a fridge, a freezer and at room temperature (Expt. 4)*

4 CONCLUSIONS

A set of simple emanation and trapping experiments have shown the extent of cross contamination that can occur with emitter composition and packaging, storage temperature and exposure time.

1. Significant contamination of desiccants (silica gel, zeolite and $CaSO_4$) was observed in unsealed containers exposed to tritiated concrete powder at room temperature and in a refrigerator. This contamination was not seen for metals and plastics.
2. Contamination of desiccants exposed to tritiated concrete was significantly reduced by storing the samples in a freezer or within sealed vials.
3. Emanation of tritium was most significant from metals (nominally 30% loss over 2 weeks storage) and to a lesser extent concretes stored at room temperature (6% loss) and in a refrigerator. Emanation from desiccants was considerably lower. Again storage in sealed containers and in a freezer effectively eliminated the potential for cross contamination.

The recommended approaches for storing samples intended for tritium analysis are :-

- Store samples in a freezer in double plastic bags to avoid cross-contamination
- Store samples in vapour tight jars or boxes
- Segregate low activity samples from high activity samples where possible

References

UNSCEAR, Sources and effects of ionizing radiation: sources and biological effects, 1982. Report to the General Assembly, with Annexes, United Scientific Committee on the Effects of Atomic Radiation. UN, New York, 1993.

Villa, M. and G. Manjon (2004) Low-level measurements of tritium in water. *Applied Radiation and Isotopes*, **61,** 319-323.

Fabre, A., et al. (2005) On the correlation between mechanical and TEM studies of the aging of palladium during tritium storage. *Journal of Nuclear Materials*, **342,** 101-107.

Evans, E.A. (1992) Guide to the Self-Decompositon of Radiochemicals., Chalfont, UK: Amersham International plc.

Harms, A.V. and S.M. Jerome (2004) Development of an organically bound tritium standard. *Applied Radiation and Isotopes*, **61,** 389-393.

Ware, A. and R.W. Allott (1999) Review of methods for the analysis of total tritium and organically bound tritium. National Compliance and Assessment Service: Lancaster.

Pointurier, F., N. Baglan, and G. Alanic (2004) A method for the determination of low-level organic-bound tritium activities in environmental samples. *Applied Radiation and Isotopes*, **61,** 293-298.

Warwick, P.E., I.W. Croudace, and A.G. Howard (1999) Improved technique for the routine determination of tritiated water in aqueous samples. *Analytica Chimica Acta*, **382,** 225-231.

Oya, Y., et al. (2001) Tritium contamination and decontamination study on materials for ITER remote handling equipment. *Fusion Engineering and Design*, **55,** 449-455.

RADIONUCLIDE RECORDING LEVELS AND PRIORITISATION OF CHEMICAL / RADIOCHEMICAL ANALYSES OF MAGNOX WASTES FOR NIREX COMPLIANCE

C. Kirby[1], D.J. Hebditch[2] and R.E. Streatfield[2]

[1]Engineering & Technology, Project Services Ltd, Risley, Warrington, WA3 6AS
[2]Engineering & Technical Services, Magnox Electric Ltd, Berkeley Centre, GL13 9PB

1. INTRODUCTION

One of the UK Nirex requirements for the safe management of radioactive waste packages is to produce a radioactive inventory. Previously the Threshold Recording Level (TRL) method was used. This was a single set of action levels for the different radionuclides. Those estimated to be below the TRL required no inventory data to be recorded. This developed through Threshold Detailed Recording Levels into a methodology using Nirex Guidance Quantities (GQs). The GQs act as a boundary between the types of inventory data that should be provided. Nirex identify 112 radionuclides requiring examination. When a radionuclide is assessed to be present at a concentration greater than its GQ, then it will need to be subject to detailed determination at the individual waste package level. This may be by chemical/radiochemical analysis or theoretical calculation based on fission product / actinide contamination, neutron activation or isotopic ratios based on these (fingerprints). If the assessed concentration is lower than the GQ, the simple bounding concentrations used in the significance test should be adequate for data recording purposes. Derivation of the GQs considered safety issues for waste transport and operational and post-closure repository performance. Separate study was given to which type of package was used for the waste. Situations involving single packages and collections of packages led to two lists of GQs for each package type. Radionuclides that are expected to be below the GQs for both a single waste package and a collection of packages should have an upper limit reported. Those above the GQ should have a best estimate plus uncertainty reported.[1] Further Nirex GQ guidance describes how the inventory needs to be decay corrected for a number of specified intervals to allow for the in-growth of daughter radionuclides.[2] This supports the repository safety case.

2. METHOD

Radionuclide volume concentrations for 86 waste streams for the representative Magnox reactor sites of Bradwell (BWA), Hunterston A (HNA) and Hinkley Point A (HPA) were taken from the Nirex 2001 United Kingdom Radioactive Waste Inventory and compared to Nirex GQ values for individual waste packages and package ensembles to produce a list of reportable radionuclides.[3] The streams studied were solid, e.g. fuel element debris (FED), gravel and nimonic springs, and mobile, e.g. ion exchange (IX) material, sludge and desiccant, operational intermediate level wastes (ILW) stored at these sites.[4] They were chosen to comprise some waste streams that would be contaminated by fission products (FP) and some that would include a range of activation products to maximise the inclusion of radionuclides.

The following methodology was used. The Guidance Quantity values depend on the type of waste package being considered. The waste streams used in this report are most likely to be packaged in either Nirex boxes/drums ($3m^3$) or drums (500 litres). These would be transported within a Reusable Shielded Transport Container (RSTC) as Type B packages as defined in the IAEA Transport Regulations.[5] Therefore the GQ lists for Type B packages have been used in this work for comparisons. They were taken for individual packages (GQ_P) and package collections (GQ_C) from Tables 2 and 3 of reference.[1]

Separate specific activity values were calculated for comparison with GQ_P and GQ_C. The specific activity for comparison with GQ_C is the average specific activity for a full waste stream taking into account uncertainties. The activity for comparison with GQ_P must reflect the potential variability of the activity in individual waste packages. They could be higher or lower than the average specific activity depending on heterogeneity in the waste. An estimate of this variability is given in the Nirex inventory, typically a factor of 10 higher or lower than the mean value.

The activity values, for comparison with the GQ_C for collections of waste packages, were calculated in the following way. Mean unconditioned activity concentration values for individual waste streams were taken from the 2001 UK Inventory.[3] These were increased by the uncertainty in the activity values and decay corrected to 01/04/2008 consistent with previous work. The product of the conditioning factor (the ratio of packaged volume to stored unpackaged volume) and the usable waste package volumes were applied to the unconditioned activity concentrations to determine the mean package activity. [3] The number of packages produced considered both current waste stocks and future estimated arisings. The activity values to be compared to the GQ_P for individual waste packages were calculated as follows. The activity values used for GQ_C which already include factors such as waste conditioning, package volumes, and activity estimate uncertainties were increased by the range quoted in the Nirex inventory to account for package variability.

In this analysis it was assumed that none of the waste streams would be mixed even though physically it would probably happen. This was done to give the conservative case that an individual radionuclide would exceed the relevant Nirex GQ. If streams with different sets of radionuclides and concentrations were mixed, the average for a given radionuclide may fall below the given GQ. Where FISPIN ratios have been used they were calculated for an average Magnox fuel burn-up, i.e. 5GW(t)d t^{-1}, and rating 3MW(t) t^{-1} followed by a cooling period of 5 years. All ILW will have been stored on site for at least this amount of time before transport and disposal. The cooling time allows very short-lived nuclides irrelevant to waste disposal to decay. Nirex scoping calculations used in the derivation of the GQ values showed that nuclides with half-lives of less than 10 days could be eliminated from consideration.

3. RESULTS

This analysis produced a list of radionuclides for each waste stream that need to be reported with uncertainties because current estimates in the Nirex inventory of the radionuclides exceed their GQs.[4] This information is summarised in Table 1 where it is broken down first into mobile and solid wastes and then into packages and their ensembles. A total of 43 radionuclides in solid waste streams and 40 radionuclides in mobile waste streams were found whose activity concentration in the final waste package exceeded either the package GQ_P or collection GQ_C. A final superset of 49 radionuclides is shown including radionuclides from both lists. Examination of the superset showed that most of these nuclides are either routinely or easily, measured or estimated leaving just 15 nuclides which would require more detailed individual consideration. These 15 nuclides are shown in Table 2 together with their production route, decay mode and half life. Prioritisation of development of new chemical/ radiochemical analysis methods was made given that such analysis generally should be performed only if the accuracy of measurements is expected to be better than that of theoretical calculation. The individual radionuclides are discussed below.

Beryllium-10: Be-10 is an activation product of Be-9. It was identified in several mobile and solid waste streams but exceeded the GQ only in fuel channel components (FCC) at HNA due to the natural Be content of the Zr D-bar fuel component. Laboratories within Magnox Electric do not currently analyse waste for Be-10 therefore the 'Best Estimate' activity in the Nirex

Table 1 : *Radionuclides that may need reporting for Solid and Mobile Waste Streams*

Solid Waste Streams	
Collection (42)	H3, Be10, C14, Cl36, Ca41, Mn54, Fe55, Ni59, Co60, Ni63, Zn65, Se79, Sr90, Nb92, Mo93, Zr93, Tc99, Cd113m, Sn121m, Sb125, Te125m, I129, Cs134, Cs137, Pm147, Eu152, Eu154, Eu155, U234, Np237, Pu238, U238, Pu239, Pu240, Am241, Pu241, Am242m, Cm242, Pu242, Am243, Cm243, Cm244
Package (26)	H3, C14, Cl36, Mn54, Fe55, Ni59, Co60, Ni63, Sr90, Cd113m, Sn121m, Sb125, Te125m, Cs134, Cs137, Pm147, Eu152, Eu154, Eu155, U235, Pu238, Pu239, Pu240, Am241, Pu241, Cm244
All (43)	H3, Be10, C14, Cl36, Ca41, Mn54, Fe55, Ni59, Co60, Ni63, Zn65, Se79, Sr90, Nb92, Mo93, Zr93, Tc99, Cd113m, Sn121m, Sb125, Te125m, I129, Cs134, Cs137, Pm147, Eu152, Eu154, Eu155, U234, U235, Np237, Pu238, U238, Pu239, Pu240, Am241, Pu241, Am242m, Cm242, Pu242, Am243, Cm243, Cm244
Mobile Waste Streams	
Collection (40)	H3, C14, Cl36, Ca41, Fe55, Co60, Ni63, Se79, Sr90, Zr93, Tc99, Ru106, Sn121m, Sb125, Sn126, I129, Cs134, Cs135, Cs137, Ce144, Pm147, Sm151, Eu152, Eu154, Eu155, U234, U235, U236, Np237, Pu238, U238, Pu239, Pu240, Am241, Pu241, Am242m, Pu242, Am243, Cm243, Cm244
Package (23)	H3, C14, Cl36, Co60, Ni63, Sr90, Ru106, Sn121m, Sb125, Cs134, Cs137, Pm147, Eu154, Eu155, U235, Pu238, U238, Pu239, Pu240, Am241, Pu241, Cm243, Cm244
All (40)	H3, C14, Cl36, Ca41, Fe55, Co60, Ni63, Se79, Sr90, Zr93, Tc99, Ru106, Sn121m, Sb125, Sn126, I129, Cs134, Cs135, Cs137, Ce144, Pm147, Sm151, Eu152, Eu154, Eu155, U234, U235, U236, Np237, Pu238, U238, Pu239, Pu240, Am241, Pu241, Am242m, Pu242, Am243, Cm243, Cm244
Superset of all 49 Radionuclides that may need reporting	
	H3, Be10, C14, Cl36, Ca41, Mn54, Fe55, Ni59, Co60, Ni63, Zn65, Se79, Sr90, Nb92, Mo93, Zr93, Tc99, Ru106, Cd113m, Sn121m, Sb125, Te125m, Sn126, I129, Cs134, Cs135, Cs137, Ce144, Pm147, Sm151, Eu152, Eu154, Eu155, U234, U235, U236, Np237, Pu238, U238, Pu239, Pu240, Am241, Pu241, Am242m, Pu242, Am243, Cm242, Cm243, Cm244

inventory is probably based on activation calculations for the components in question. The estimates generally exceed the GQ limit by a factor of 250 times. The upper activity band (C) is 10 times the median value quoted and was used in the calculation. For the limited number of waste streams involved it would be prudent firstly to examine in detail the assumptions used in estimating the Be-10 content. Possibly another measurement of the inactive Be content of the

Zr D-bars could reduce any uncertainty on the estimate. If the estimate and the data it is based on are verifiable, then further measurements are not required.

Calcium-41: Ca-41 is produced by neutron activation of natural Ca-40. It has been found to exceed the GQ by a factor of 2 in both graphite fuel struts and desiccant from HNA (3 streams). The reported desiccant value is an upper limit probably based on trace contamination by graphite dust. In decommissioning wastes, activation of the concrete bioshield would also be expected to produce Ca-41 but these wastes streams are regarded as low level wastes in the NIREX inventory and hence GQ values do not apply. Measurements of Ca-41 can be obtained after chemical separation of Ca, which is done routinely for Ca-45 measurements. After any Ca-45 ($t_{½}$ =163 days) has decayed away, it can be measured by liquid scintillation counting. Procurement of direct standards from NPL would be required. In fresh samples, if the Ca-45 has been measured, then the Ca-41 could be estimated by comparison of activation

Table 2 : *15 Radionuclides potentially difficult to measure directly in Magnox Wastes*

	A / F †	Production Route or Fission Yield	Decay Mode †	Half Life (y)
Be10	A	Be9 (n,g)	Beta	1.60E+06 ± 0.20E+06
Ca41	A	Ca40 (n,g)	EC	1.03E+05 ± 0.04E+05
Se79	F A	0.044% Se78 (n,g)	Beta	<6.50E+04
Nb92	A	Mo92 (n,p) Nb93 (g,n)	Pos	3.50E+07 ± 0.30E+07
Mo93	A	Mo92 (n,g) Nb93 (p,n)	EC	3.50E+03 ± 0.70E+03
Zr93	F A	5.986% Zr92 (n,g)	Beta	1.53E+06 ± 0.10E+06
Tc99	F	6.074%	Beta	2.13E+05 ± 0.05E+05
Cd113m	A F	Cd112 (n,g) 0.015%	Beta	1.41E+01 ± 0.05E+01
Sn121m	A F	Sn120(n,g) 0.013%	IT/Beta	5.5E+01 ± 0.5E+01
Sn126	F	0.054%	Beta	ca 1.00E+05
Cs135	F A	6.536% Sm150 (n,g)	Beta	2.30E+06 ± 0.30E+06
Np237	F	U237 decay	Alpha	2.14E+06 ± 0.01E+06
Am242m	A	Am241 (n,g) or U238, Pu239 +n's	IT	1.41E+02 ± 0.02E+02
Pu242	F	U238, Pu239 +n's	Alpha	3.74E+05 ± 0.01E+05
Am243	F	U238, Pu239 +n's	Alpha	7.36E+03 ± 0.02E+03

† A Activation product; F Fission product; EC Electron capture; IT Internal transition; P Positron decay

calculations because both are formed from inactive Ca. Further development of an analytical technique therefore does not seem appropriate.

Selenium-79: Se-79 is a low yield FP and can be produced by neutron activation of natural Se-78. At HPA and HNA, it was found to exceed the GQ in various contaminated items and in IX resin due to potential contamination by fuel products. Adequate estimates of the amount of Se-79 present were possible where full radiochemical analyses had provided information on other FPs. HNA desiccant was found to exceed the GQ most (by a factor of 25). The reported level is an upper estimate and probably reflects the possibility that volatile Se compounds might be transported to the desiccant where they concentrate in a similar way to S-35. Production by activation may be more important for this waste stream. Hence it is more difficult to predict ab initio the Se-79 concentration in desiccant. Desiccants from the other stations did not exceed the GQ and the HNA estimate may be overly conservative. Any analytical method for Se-79 would need to attain a minimum detectable activity (MDA) ≤6 Bq cm^{-3}, which is required for the GQ for the more restrictive condition of a collection of packages. Because Se-79 can be formed by fission and activation and Se chemistry could cause concentration of the element in certain wastes, estimation is difficult and development of an analytical method could be appropriate.

Niobium-92 and Molybdenum-93: Nb-92 and Mo-93 are both activation products of Mo-92. The estimates in the Nirex inventory were found to exceed the GQ levels in waste streams containing Magnox FED at HPA by 6 times and in FCC at HNA by 50 times (Mo-93 only). These radionuclides are not currently measured within Magnox. Their respective GQs for collections of packages are equivalent to 0.04 and 200 Bq cm^{-3} in the conditioned waste. Nb-92 is a gamma emitter but it is unlikely that the detection limit required could be achieved by direct counting in the presence of the other gamma nuclides. Some chemical separation is needed if it were to be measured. Mo-93 decays by positron emission and could possibly be measured by liquid scintillation counting after chemical separation. Both radionuclides are non-volatile and are likely to be localised in the solid objects where produced. There is no realistic method by which they could be concentrated in a waste stream, so activation calculations should provide a good estimate of their activity if the trace, precursor levels of Mo and Zr are well known. It is unlikely that further analytical development for these radionuclides would be cost effective.

Zirconium-93: The main production route for Zr-93 at Magnox sites is activation of Zr-92 in natural Zr. FCC at HNA exceed the GQ level (200 Bq cm^{-3}) by 500 times. This is consistent with activation of Zr D-rings. Magnox FED waste streams at HPA are estimated to exceed the GQ level by 5 times. This is consistent with the activation of a small amount of zirconium alloy contained in the waste. Zr-93 can also be produced by fission. Although the direct fission yield for Zr-93 is similar to that for Cs-137, its much longer half life means that its specific activity in waste under the conditions considered is small by comparison (50×10^3 lower). Fission as a source of Zr-93 would only be important where there is a significant contamination by FPs as indicated by a very high Cs-137 or alpha presence. Sand from the sand pressure filters (SPF) at HPA is an example and exceeds the GQ level by 1.5 times. The quoted uncertainty on the Zr-93 activity (and all the activities in this waste stream) is a factor of 100, which is why it exceeded the GQ. Efforts to reduce the uncertainty are needed by, for example, measurement of other activities. This may remove the need to report Zr-93 in this stream. Estimates of Zr-93 can thus be made by calculation where it is associated with individual items containing Zr (activation) or significant fuel contamination (FISPIN code). It is unlikely that development of an analytical procedure for this radioisotope would be cost effective.

Technetium-99: Tc-99 is a reasonably high-yield (6%) FP (similar to Cs-137). The ratio of activities Cs-137 to Tc-99 is calculated (FISPIN) to be 6,000 to 1. Of the selected waste streams in the UK Nirex Inventory, 18 were found to exceed the GQ level for Tc-99 including HPA SPF (high uncertainty), FED and miscellaneous contaminated items (MCI), and HNA FCC, MCI and IX resin. Tc is chemically different to Cs because it will form anionic species (the pertechnetate anion, TcO_4^-) whereas Cs will only form cations. Tc-99 will not be retained by the standard resins used to remove Cs. Cs-137 will therefore not be a good marker for Tc-

99 and could not be used reliably for fingerprinting wet wastes. For this reason it may thus be difficult to estimate by calculation the Tc-99 content of wet waste streams such as the sand from the filters and IX resins. A radiochemical analysis would confirm the current estimates in the inventory based on FISPIN. Tc-99 is not currently routinely analysed in Magnox wastes but there has been extensive investigation of Tc-99 at Sellafield regarding its behaviour during reprocessing and liquid waste treatment. Analysis of Tc-99 has been carried out in a wide range of environmental samples such as groundwater, oils and sludge. Extending these methods to Magnox wastes would require some development.

Cadmium-113m: Cd-113m is produced by activation of Cd-112 (n,γ) and also has a very low fission yield. The predicted ratio of Cs-137/Cd-113m is 70,000 to 1. It exceeds the GQ value in 13 waste streams comprising FCC from HNA and in FED from BWA and HPA. It is not currently analysed in Magnox wastes. The reported values in the inventory are best estimates based on activation of impurities. The maximum by which Cd-113m exceeds the GQ value is 20 times. It would take a further ~60 years decay before it fell below the GQ. It would thus continue to need reporting over the expected disposal period even though Cd-113m has the shortest half life of those nuclides identified for further study (14.1y). It is a pure beta emitter and hence any analytical method would require its complete chemical separation before liquid scintillation counting. These waste streams are dominated by activation products. Therefore this radioisotope can be estimated by activation calculations making it unnecessary to develop an analytical method.

Tin-121m: The main source of Sn-121m is neutron activation of natural tin, Sn-120, but it also has a low fission yield. The predicted ratio of Cs-137 to Sn-121m is 6,000 to 1. Sn-121m is not currently analysed in Magnox wastes. The estimates in the Nirex inventory exceeded its GQ limit in the following waste streams; IX resins at BWA and HPA and FCC at HNA. IX resins are contaminated mainly by FPs from the fuel storage pond whereas the FCC activity arises mainly by activation of steel and Zr. Reasonable estimates of Sn-121m in these waste streams should be possible using FISPIN or activation calculations as appropriate especially when uncertainties in the common fingerprint activities will be reduced by further analysis of the waste streams.

Tin-126: Sn-126 is a FP. The predicted ratio of Cs-137 to Sn-126 is 100,000 to 1. It exceeded the GQ in IX resin wastes from BWA and HPA. IX resins tend to be dominated by FPs. Although Sn is less soluble chemically than Cs, the small amounts of Sn produced by fission could follow Cs. In the Nirex inventory Cs-137 has been taken as an adequate marker for this isotope and estimated using calculated fission ratios.

Caesium-135: Cs-135 is a high yield FP. It was found to exceed the GQ level in IX resins and SPF media. It behaves in the same way as Cs-137 and can be estimated easily using FISPIN calculations. There is no need to produce an analytical method for this isotope.

Neptunium-237: Np-237 has two production routes. The first is rapid β decay of U-237 produced by nuclear reactions in the fuel. The second route is β decay of Pu-241 to Am-241, which decays losing an α particle to Np-237. This second route will lead to the in-growth of Np-237 in the waste during storage. Np-237 will only be present in waste contaminated with actinides. This is demonstrated by the waste streams that were found to exceed the GQ level. They were IX resin and sludge from BWA and SPF contents from HPA. Adequate estimates of Np-237 can be made by ratio to Cs-137 determined by FISPIN in the first instance and then refined by taking account of the accumulation and chemistry of other transuranics already measured. At higher concentrations, it can also be measured by γ spectrometry of the short-lived daughter Pa-233. In-growth was included for Np-237, where data were decayed to the common date of 2008. Development of an analysis technique is unnecessary.

Americium-242m: Am-242m is produced in fuel by the activation of Am-241. Its estimated inventory value exceeds the GQ level for HPA and HNA IX resins, HPA SPF sand, BWA sludge and the MCI streams at HNA and HPA. These are wastes streams with FP contamination. Am-242m is not analysed in Magnox wastes but has been observed as interference in the alpha spectrum of Cm-242 from BWA sludge. Am-241 is regularly analysed in waste and hence combined with FISPIN calculations could be used to estimate Am-242m in the normal way. The expected ratio of Am-241 to Am-242m in typical Magnox fuel contamination is approximately 200 to1.

Plutonium-242 and Americium-243: Pu-242 and Am-243 are produced in fuel by multiple nuclear reactions. They therefore appear in items contaminated by fuel. They were found to exceed the GQ limit in HNA and HPA MCI and HPA SPF waste (high uncertainty). In addition in IX resins at HPA and HNA, Am-243 was above the GQ. Neither of these radionuclides are currently analysed in Magnox wastes because they are used as yield tracers in other analyses. To measure these two radionuclides, it is possible to simply repeat the current analyses for Pu and Am; with and without tracers. No development work should be required. It has been possible to use these isotopes as tracers because the amount present (in terms of activity) is very low. FISPIN predicts the following radioisotope ratios in fresh waste; Am-241 to Am-234 of 111 to 1 and Pu-239/-240 to Pu-242 of 2,500 to 1. At these activity levels, it may be more accurate to estimate the activities rather than measure them.

4. DISCUSSION

Usually, the radionuclide inventory of a waste stream is obtained through a combination of measurements and estimates of unmeasured radionuclides. A full range of measurements is carried out on a number of samples of the waste and then ratios are calculated. The manner in which the unmeasured radioisotope fingerprint ratios are derived varies. Fission products have been estimated using a suitable theoretical model and computer code of nuclear reactions taking place in the fuel, for example FISPIN, using usually a typical average irradiation history for the reactor in question. Similarly the activation nuclides have been estimated by calculation using the neutron flux history of the specific waste and a knowledge or best estimate of the precursor elemental concentrations in the irradiated material. A simple fingerprint is produced by estimating the ratio of fission product activities to Cs-137 and the activation products to Co-60. The inventory for a specific waste package is then estimated by measuring the Cs-137 and Co-60 dose rates. This approach may also be modified to take account of any differential concentration / dilution processes taking place during the lifetime of the waste after the irradiation has ceased using a measured nuclide with similar chemistry.

The activities of the fifteen nuclides of interest identified above were determined for the 2001 UK Nirex Inventory in this way. The question that this report addresses could be re-phrased as "Is the current method of estimation adequate?" This is discussed in the section above and the findings are summarised in Table 3. It was found that for the most part the current methods of estimation by calculation were adequate. The radionuclides are formed either by activation or pure fission. There were chemically similar nuclides to which they could be correlated.

Table 3 : *Summary of Findings*

Nuclide	Proposed Method of Determination	Further Work Required?
Be10	Activation calculations	No
Ca41	Activation calculations	No
Se79	Possible problem with calculational route owing to volatility and accumulation.	Develop analytical method. Especially for desiccants.
Nb92	Activation calculations or gamma spectrometry where possible	No
Mo93	Activation calculations	No
Zr93	Activation or fission calculations as appropriate for each waste stream	No
Tc99		Apply external methods to Magnox waste streams
Cd113m	Activation calculations	No
Sn121m	Activation or fission calculations as appropriate	No
Sn126	Fission calculations	No
Cs135	Fission calculations	No
Np237	Fission calculations	No
Am242m	Fission calculations	No
Pu242	Fission calculations	Modify existing plutonium procedure
Am243	Fission calculations	Modify existing americium procedure

Some nuclides are formed both by activation and fission. In wastes streams in which these are important, their origin is dominated by a single formation method.

Pu-242 and Am-243 are not currently measured but need only small change of an existing procedure. This may be worthwhile when expected to be present in significant amounts.

Development of a new analytical method was proposed for Se-79. Not only can Se-79 be produced by activation and fission but its chemistry also plays an important role. In desiccant, its concentration could be enhanced due to its volatility. It was also suggested that current methods for Tc-99 used at Sellafield be extended to Magnox wastes that have a high FP contamination and failing this a new analytical method be developed. R&D work has commenced to design, develop, realise and deliver radioanalytical methods for Se-79 and Tc-99, which include literature survey, consideration of matrices/environments in which these radionuclides might be expected to be concentrated, allowance for minimum detectable activity and decontamination factor (DF) of typical interfering fission products, laboratory based radiochemistry development, replicate studies and validation, statement of the minimum detectable activity associated with the analytical methodology, etc.[6]

The disposal criteria (GQs) are not the only reasons why it may be necessary to extend our analytical capabilities. There are nuclides that although they can be measured, require better limits of detection than those currently available. Table 4 gives 3 examples, Nb-94, Ag-108m and Ho-166m. These are important long lived gamma emitters that will be important for dose assessment in steel structures during decommissioning when Co-60 has decayed away. Work on better analytical measurements has been carried out.[7] There is also work underway to increase the accuracy of the calculation routes. Improving our knowledge of the impurity levels (precursors) in materials for irradiation would lead to better estimates of activation products. For example, the concentration of lithium in steels has recently been measured to give improved estimates of tritium production.[8]

5. CONCLUSIONS

The introduction of NIREX GQs for packaging and disposal of ILW has resulted in the need to analyse or calculate the specific activities of radionuclides that had not been important historically. This paper identifies those nuclides that are likely to exceed the GQ in a range of Magnox waste streams and considers whether adequate procedures exist for all. For those

Table 4 : *Examples of Radionuclides benefiting from improved limits of detection*

	A / F †	Production Route or Fission Yield	Decay Mode †	Half Life (y)	Uncertainty
Nb-94	A	Nb-93 (n,g)	Beta	2.03E+04	1.60E+03
Ag-108m	A	Ag-107 (n,g)	EC IT	4.18E+02	1.50E+01
Ho-166m	A	Ho-165 (n,g)	Beta	1.20E+03	1.80E+02

† A Activation product; F Fission product; etc.

radionuclides that cannot be determined by robust chemical/radiochemical analysis techniques, the suitability of other methods of estimation is considered and the benefit of development of new analysis methods assessed. A total of 86 mobile and solid Magnox operational ILWs were compared to the UK Nirex Guidance Quantities (GQs) for Type B packages and ensembles using data from the 2001 UK Nirex Inventory. This assessment took into account expected waste processing routes and uncertainties in specific activities.

- From the 86 streams selected as representative of Magnox reactor operational solid and mobile intermediate level wastes, 49 radionuclides were identified which are subject to detailed determination under UK Nirex requirements for radioactive ILW packages.
- Examination of these reportable radionuclides showed that most are either routinely or easily, measured or theoretically calculated, leaving just 15 requiring detailed investigation.
- Assessment and judgement of the 15 more unusual radionuclides, associated waste streams and related uncertainties was used to prioritise the benefit of development of new chemical/radiochemical analysis methods as compared to further theoretical calculations.
- Clear benefit for development of new chemical/radiochemical methods for analysis was found for only two radionuclides, Se-79 and Tc-99, and this work is now underway.
- Se-79 may concentrate in some waste streams, for example reactor desiccant, making an *ab initio* estimate difficult and a new analytical method would remove this difficulty.
- Extension of the Tc-99 analysis method from environmental samples at Sellafield to Magnox wastes contaminated with fission and activation products, for example IX resin, is required and failing this a new analysis method is beneficial.
- Development work of a complementary nature has been undertaken to improve the limits of detection of several fission and precursor activation products comprising Nb-94, Ag-108m and Ho-166m (long-lived γ-emitters) and Li as precursor to H-3.

6. REFERENCES

1 J. Jowett, Customer Guidance on the Requirements for Waste Package Radionuclide Inventories, NIREX, (SERCO), Report No. A/R/PSEG/04441, Draft 3, December 2001.
2 J. Jowett, Customer Guidance on the Requirements for Waste Package Radionuclide Inventories, NIREX, (SERCO), Technical Note, Doc. No. 408414v1, Rev: 1, August 2004.
3 DEFRA and Nirex, The 2001 United Kingdom Radioactive Waste Inventory. DEFRA and UK Nirex Ltd, DEFRA/RAS/02.006 and Nirex Report N/044, 2002.
4 C. Kirby, Analytical Systems for Nirex Compliance, Project Services, Report No. E&T/REP/GEN/1530/05, Issue 2, August 2005.
5 IAEA, Regulations for the Safe Transport of Radioactive Material, IAEA Safety Standards Series, Report TS-R-1 (ST-1 Revised), 1996.
6 Personal communication, Professor P. Warwick, Head, Department of Chemistry, Professor of Environmental Radiochemistry, Loughborough University, U.K., August 2006.
7 A. King, Ph.D. Thesis, Analysis of Long-Lived Radionuclides in Decommissioning Waste, Loughborough University, PhD Summary, BNFL NSTS/GEN/EAN/0237/03, February 2004.
8 W.A. Westall et al., Determination of ^{3}H and Li in Magnox reactor steels, Environmental Radiochemical Analysis, RSC 10th Int. Symposium, Oxford, U.K.,13-15 September, 2006.

Acknowledgment - The authors gratefully acknowledge support and permission to publish from the NDA and Magnox Electric Limited.

APPLICATION OF THE RADIOLOGICAL HAZARD POTENTIAL (RHP) TO RADIONUCLIDES IN MAGNOX REACTOR DECOMMISSIONING

R.E. Streatfield, D.J. Hebditch and W.H.R. Hudd

Magnox Electric Ltd, Berkeley Centre, Berkeley, Gloucestershire, GL13 9PB.

1 INTRODUCTION

UK Government Policy requires the systematic and progressive reduction of hazard during the waste management and decommissioning of nuclear plant. Metrics were developed to enable progress to be monitored and quantified for the conversion of radioactive materials to more passive forms. This would demonstrate to stakeholders that on-going hazard reduction was actually occurring and that one hazard was not being replaced with another similar one. The paper explains how the hazard is quantified in terms of the radionuclides present, the potential for dispersion and the degree of control required to manage the hazard. Numerical factors are given with examples of application to diverse waste streams.

Magnox Reactor waste streams include a wide range of materials such as ion exchange (IX) resins, sludge, Magnox fuel element debris (FED), reactor graphite and carbon and stainless steels. Some wastes will exhibit heterogeneity and for many construction materials such as steels there will be radionuclides present which are neutron activation products of trace impurities and were un-quantified at the time of manufacture.

The paper considers how the updated metric, named Radiological Hazard Potential (RHP), can be summated for the radioactive waste streams on a particular site and used to help set priorities for future waste management activities. Uncertainties in RHP values may be utilised to prioritise further sampling and radio-analytical measurements. The paper outlines the difficulties in dealing with hazards posed by non-radioactive materials, and some cautionary advice is given on the correct application of the RHP, particularly in avoiding its use as the sole criterion for establishing work priorities.

2 BACKGROUND

In the final conclusions of Command 2919, it is stated that "In considering proposals for decommissioning nuclear plant put forward by the operators, the HSE NII will assess them to ensure that the proposals assure the safety of the site at all times, and that the hazards presented by the plant (or site in the case of nuclear power stations) are reduced in a systematic and progressive way".[1] Cm 2919 was first published in July 1995, and was endorsed by other groups, and responded to by different sections of the nuclear industry.

An early issue was the difference between risk and hazard. A hazard is the potential for the intrinsic properties or the disposition of a material to cause harm. Risk is the chance that someone will be adversely affected. The concept of hazard potential is a measure of the harm that could occur or be caused by the material in the form that it is currently in. The hazard from radioactive waste is due to its intrinsic properties such as radioactivity and possible biochemical toxicity. The hazard will only be reduced by radioactive decay, whereas the hazard potential can be reduced by retrieving, immobilising and containing the waste. This reduces the risk of the hazard realising its potential.

Early work in this area was carried out in 2001 by the UKAEA, and the "Hazard Reduction Index" illustrated the extent of hazard reduction to be achieved by UKAEA through their site strategies and plans. This work was shared with BNFL and other UK nuclear organisations in 2001, and since then BNFL and UKAEA have further developed the concept and published a proposed Waste Conversion Index (WCI) in September 2003.[2]

A number of observations were made by the various review groups as follows.

- The WCI was a useful parameter for direct comparison between waste streams.
- Stakeholders could gauge progress in hazard reduction for a given site.
- Confidence needed to be established in the methodology, including use of the ICRP Specific Instantaneous Toxic Potential concept and the numerical scale proposed.
- The assumptions underpinning the methodology, provenance of the data and factors used would need to be made explicit and transparent.
- Buy-in from a wide range of stakeholders was seen as essential.
- The designation of some materials as potential wastes would be continuously examined, and the WCI must be capable of application to re-designated materials.

The WCI methodology was then developed nationally by a group drawn largely from the nuclear industry and including the MOD, regulators and academics. Examination showed that the factors used, while giving consistent answers, could not be made transparent enough to gain full stakeholder acceptance. Work then focussed on a methodology to meet the transparency criterion. The title 'index' carried an unjustifiable expectation of mathematical precision. The 'hazard potential', was then addressed rather than the 'hazard'. Accordingly, the final choice was to use the term Hazard Indicator (HI), and require a clear statement of purpose, lucid explanation of the factors used for calculation and justification of the values attributed to these factors.

Historically the HI and its precursors, have mainly been applied to Intermediate and High Level Waste (ILW and HLW). The HI can be applied to Low Level Waste (LLW) and large decommissioning areas such as rubble, steelwork and contaminated land. As UK LLW is routinely disposed of to Drigg, the value for LLW is in predicting the hazard retained on a site that is yet to be fully decommissioned.

Since the formation of the Nuclear Decommissioning Authority (NDA) there has been an emphasis on a demonstrable process for prioritisation of remediation work on UK decommissioning sites. As part of this process, the HI has now been re-named the Radiological Hazard Potential (RHP) and is one of five metrics which are combined numerically to produce a prioritisation metric, the Safety and Environmental Detriment (SED), see below.[3] This paper relates only to the RHP. It is a quantitative measure of the potential for a material or plant item to cause harm, but does not address the risk of that harm occurring and has universal application irrespective of facility type.

3 CURRENT STRATEGY FOR WASTES

The potential for a radioactive material to cause harm is generally highest when stored in its un-treated state. Many of the wastes associated with Magnox power stations have been stored for many decades and there is the possibility that the hazard represented by a waste can increase with time. Wastes can range from potentially mobile materials such as sludges and ion exchange resins to discrete metal items. The potential to cause harm is reduced by retrieving and conditioning the waste into a more passive form suitable for long term storage. The current reference material to achieve this is a cement matrix, in which the detailed cement formulation has been adjusted to provide compatibility with the different waste materials. Wastes are immobilised in one of a range of Nirex compatible stainless steel containers, and these containers would be stored on-site in an appropriate location until a disposal facility becomes available.

The proposed uses of the RHP can therefore be summarised as:

- To track the reduction in the hazard potential of a waste from one storage condition (e.g. vault or tank) to another such as immobilised in cement in a stainless steel container.
- To track the reduction in the associated hazard potential within individual storage facilities, sites and across the UK.
- To compare scenarios, and prioritise decommissioning activities as part of the planning process, on the basis of their effectiveness in reducing hazard potential.

4 TIMESCALE

The RHP excludes the very short term (to about 3 months after shutdown) and the long-term activities concerned with transport to the repository, disposal and the evolution of wasteforms on geological timescales. The emphasis has therefore been to develop a measure that will reflect the amounts of nuclear materials in storage and the impact of decommissioning/cleanup activities in the medium term (10 to 30 years) during which time wastes will be converted into forms suitable for interim storage.

Calculations were made for the initial and final states of the waste, and also year by year to show the impact of radioactive decay. Modelling of any short-term increases in hazard potential arising from retrieval, immobilisation and store management operations is outside the scope of the RHP, which is intended to be a measure of progress towards passively safe storage rather than a continuous "hazard monitor". The RHP is normally calculated either annually or upon work-stream completion.

5 TECHNICAL DETAILS

The RHP is a relative rather than an absolute measure, allowing comparison between different radioactive materials and between materials stored in different forms at different locations. The inherent potential to cause harm is influenced by:

- The radiological inventory (specific activity and radionuclide composition).
- The physical state or form of the material – gas, liquid, solid etc.
- The ease or difficulty of storage using a specified storage mode taking into account the chemical, physical and radiological properties of the material

The RHP attempts to address these by the use of the following factors:

5.1 Radiological Inventory

Of particular relevance is the inherent hazard represented by the radioactive inventory of the waste and is the summed product of the activity (Bq) of each radionuclide present and its corresponding Specific Instantaneous Toxic Potential (SITP). This can be adjusted for decay for longer timescales and it is usual to employ the full radionuclide inventory for waste streams as sampled and quantified by Magnox Electric Ltd and reported in the U.K. Radioactive Waste Inventory.[4]

Consequently the parameter used is the Specific Instantaneous Toxic Potential (SITP) of the material, which takes account of the specific nature of individual radionuclides in terms of their half life, persistence in the body etc, and calculates a theoretical dose which is a better measure of the potential harm to man. SITP values are available for both ingestion and inhalation pathways. The dose/risk relationships were derived by the International Committee on Radiological Protection (ICRP) and are subject to regular review.[4,5]

The SITP is a quantity derived from the Annual Limit on Intake (ALI), an internationally accepted concept that has been acknowledged by the Government's Radioactive Waste Management Committee (RWMAC) as a valid method of establishing equivalent hazards of different waste types. The ALI is a derived limit for the permissible amount of radioactive material taken into the body of an adult radiation worker by inhalation or ingestion in a year. The ALI is the smaller value of intake of a given radionuclide in a year by the reference man that would result in either a committed effective dose equivalent of 0.05 Sv or 0.5 Sv to any individual organ or tissue.

For ingestion the SITP is defined as the volume of water required to dilute a material to a concentration that would be safe to drink if a population were to use the mixture as its sole source of water. The derivation of SITP values is as follows:

For most radionuclides and most waste streams the SITP relates to the ingestion pathway as the inhalation pathway is not a realistic possibility, for example for a large activated solid with no loose contamination, where the adult ingestion committed effective dose coefficient (e(50) Sv Bq^{-1}) value for inhalation is less than that for ingestion. The adult ingestion committed effective dose coefficient (e(50) Sv Bq^{-1}) for uptake from a single radionuclide over 50 years, as published by the ICRP, is converted to the (ALI) (Bq y^{-1}) which is the quantity which would give rise to an annual dose of 1 mSv for an adult member of the public.[5,6] The quantity of water drunk by a standard man is given by the NRPB (now the Health Protection Agency) as 0.6 m^3 y^{-1}, and therefore in accordance with the above definition the SITP is 0.6/ALI m^3 TBq.[7]

Where several e(50) values are given depending upon the solubility class of the radionuclide and the speed of metabolism in the body, the bounding or highest value in any solubility class is used. For some radionuclides and a few waste streams, it is possible that following loss of control, inhalation could provide a significant exposure pathway relative to the ingestion accounted for in the SITP. This would be a waste stream that is in the form of a gas, fine particulates or an unstable material (e.g. sodium) which could convert from a solid to fine particulate.

In this case if the e(50) for inhalation is greater than the e(50) for ingestion, (this ratio is referred to as the aerial correction factor and can vary from 1 up to 500) the former should be used to modify the SITP as follows:

SITP(modified) = SITP(ingestion) x e(50)(inhalation)/ e(50)(ingestion)

5.2 Form Factor

The form factor of the RHP represents the physical state of the material *e.g.* gas, liquid, or solid, the implied mobility and hence the potential for dispersion and subsequent ability to cause harm. The form factor addresses the amount of material that would be released from the bulk material if containment was lost completely for a short period (e.g. 1 day). In this situation, a gas or liquid might be expected to disperse quickly, but a solid block of glass would remain immobile. The form factor does not consider the likelihood of any containment loss but is intended to reflect the consequences in the event that such a situation occurs *i.e.* it represents the potential of a material to do harm, rather than the likelihood of such harm occurring.

At one end of the range of form factor values, liquids and gases are assumed to be completely released and given a form factor of 1. At the other end of the scale, a monolithic solid such as a container of vitrified HLW or a clean, activated component weighing several tons would release virtually no material and is conservatively assigned a form factor of 10^{-6} which is a release of activity of one part per million (ppm).

Between these two extremes are waste streams with the following characteristics:

- Sludges/flocs will not disperse as completely as liquids, and if data are available for the separate solid and liquid components it is recommended that separate form factors be used for each. If inventory data are only available for the sludge or floc as a whole a form factor of 1 is used for a high water/solid ratio and 0.1 for a low water/solid ratio with a substantial solids burden.
- Few radioactive materials are stored in powder form, and application has been dominated by Pu and U oxide powders. These are dense materials and while fine particles (<10 μm) might be expected to be dispersed following a release, larger particles would be relatively immobile. Experimental data used in safety cases to model airborne releases from powder spillages indicate that 10% release is a reasonable upper limit, and therefore a form factor of 0.1 is assigned to powders.
- Loose contamination on the surface of solid objects should be treated as a powder, and as the form factor for a powder is much higher than that for the solid it is associated with (0.1 compared with 10^{-6}), the results will be sensitive to assumptions made about the inventory of the loose contamination. Such assumptions must be explicit and defensible. The bulk item that has been contaminated would then be treated independently. Alternatively, a combined factor for powder plus solid can be derived.
- Discrete solid items (e.g. pellets, miscellaneous solid items) are assigned a value of 10^{-5} to differentiate their behaviour from more massive solid objects.

In deriving these conservative form factors it is assumed that more material would be dispersed than would probably be the case. This also extends to materials that could change spontaneously and rapidly to a more mobile state if control were lost, such as uranium hexafluoride, sodium metal and uranium hydride. In these cases the more mobile state should be assumed when assigning the form factor. It is important to record the assumptions behind the form factor such as the size of the solids assumed.

Form factors are only quoted to one significant figure to avoid conveying the impression of excessive precision, see Table 1.

Table 1 *: Summary of Form Factor Values*

Physical Form	Form Factor
Gases, liquids, watery sludges and flocs	1
Other sludges	0.1
Powders and loose contamination	0.1
Discrete solids	0.00001
Monolithic solids	0.000001

5.3 Control Factor

The combination of physical, chemical and radiological properties of radioactive materials varies widely, and this is addressed with a factor which measures the difficulty in storing a given material. This is length of time that it could be left with no monitoring or other intervention, whilst retaining confidence that containment would be maintained and the material would remain under control. A long time period will indicate an easily stored material, while a short time period will indicate one that is more difficult to store. The time period selected will depend not only on the properties of the material, but also on the mode of storage. The examples below demonstrate the range of differing time intervals that the control factor caters for, and how these may change with storage regime.

- Radioactive decay of highly active liquid wastes generates substantial heat and the storage tanks are continuously cooled by systems with multiple levels of redundancy. Loss of the cooling systems would eventually result in the liquids boiling and control being lost, and a high degree of control is therefore required.
- Other wastes generate H_2 and loss of ventilation may cause an explosive atmosphere. Control systems are needed to ensure concentrations are maintained at safe levels.
- UF_6 is highly unstable when exposed to air or water, but is passivating to the steel containers used to store it. UF_6 in a container ensures that the only concern is corrosion from the outside, and storage times for cylinders in covered stores are relatively long. A shorter control period is appropriate for storage in the open air.
- Solid materials that have been activated, but contain no surface contamination merely require physical isolation to provide shielding. Long control periods may be justified.

Knowledge of the conditions of the current storage facility is essential when determining the risk of a significant loss of containment, but should not be taken into account when

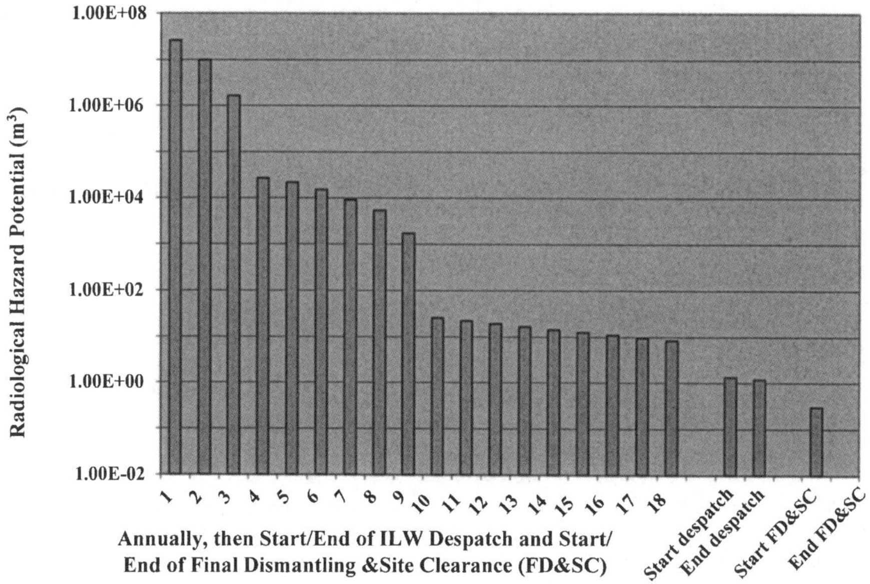

Figure 1 Total Site Radiological Hazard Indicator (incl. fuel) – Variation with Time

calculating RHP since this aims to assess the intrinsic hazard potential of the material. The control factor therefore attempts to take account of the differing intrinsic properties of radioactive materials by identifying the monitoring period for the current/proposed storage mode that a competent design engineer would propose at the design stage, as being necessary to ensure containment based on the intrinsic hazard of the material, the mode of storage, and how these might evolve. A four stage process has been developed as described below.

Stage 1 – Storage Mode

Identify the current storage mode for existing materials and a postulated storage mode for future situations. In either case, the control factor is assigned as if the facility was at the design stage, with the mode of storage starting in good condition. This applies even where the actual storage arrangements for the material have known shortcomings and a limited remaining life. Example storage modes are open ponds, open silos/tanks, closed tanks, storage building, boxes, drums etc.

Stage 2 – Eight Characteristics

The inherent physical and chemical properties of the stored material which could threaten the storage mode are considered against the following eight characteristics.

1. Is the material sufficiently heat generating to cause reaction or change of state?
2. Is the material corrosive with respect to its containment or packaging?
3. Will the material burn in air?
4. Has the material sufficient corrosion potential to change state if exposed to air or water? This will have significance if corrosion affects a significant part of the activity inventory or if corrosion could aid dispersion of the inventory.
5. Does the material produce secondary hazardous materials such as H_2 in normal storage, which would require secondary systems *e.g.* agitation or ventilation? Is the material liable to dry out and disperse as a powder?
6. Does the material react exothermically with water?
7. Does the material react exothermically with air?
8. Are other factors such as Wigner energy or criticality relevant? Is there the potential for reactivity excursions? In practice it is unlikely that either criticality or Wigner energy will affect the control factor, since the RHP relates to steady state storage rather than operations.

Stage 3 – Assign Control Factor.

With the material in the identified storage mode and reviewed against the eight potentially hazardous characteristics, the control factor for each material can be assigned using engineering judgement based on its properties and the requirements these place on the storage mode. It is the *material* being stored and its *mode* of storage, and *not* the state of the particular plant or facility, which is considered when assigning the control factor. Thus additional containment of the same type will not change the control factor, whereas a move into a different containment *mode* could have an effect.

As the purpose of the control factor is to distinguish between materials that are more difficult/relatively easy to store, the process should identify those features of the storage mode which need to remain effective in order that the material remains sufficiently under control and a bulk release cannot take place. This will involve keeping storage conditions within identified design parameters to ensure containment, and the control factor will be the time that would be required without corrective action for conditions to evolve to the point where they are no longer within these parameters.

As an example, consider a covered pond used to store corrodible spent fuel with storage characteristics involving water temperature, pH, and the presence of a covering layer of water. A failure of any of these features would prejudice the ongoing containment of the stored fuel, and consequently the design engineer would propose monitoring to ensure these parameters remained within specification. The rate of heat release from the fuel would determine the water temperature and the potential for rapid corrosion, a change of pH could also affect the rate of corrosion and hence the release of activity, and the presence of water would be assured unless it was removed by evaporation. It may be judged that the temperature would take days to breach desired conditions, the pH would take weeks and the water loss by evaporation months. The control factor would therefore be measured in days. Other possible means of loss of control such as those associated with the age or specific features of the actual facility, or with aspects of the idealised mode of storage that cannot be classed as storage features should not be considered in this example of the control factor.

Stage 4 – Rounded Control Factor.

Based on the limiting time without corrective action for conditions to evolve to the point where they could prejudice the ongoing containment of the stored material, the appropriate control factor should be assigned using the table below. Exact control factors based on time have been avoided because of the unjustified precision inferred, and rounded decades of time in hours have been used, see Table 2.

Table 2 *: Control Factor Values*

Control Time	**Typical Factor**	**Rounded Decade Control Factor**
Hours	1	1
Days	24	10
Weeks	168	100
Months	730	1,000
Years	8,760	10,000
Decades	87,600	100,000

It is important that the assumptions made in each of the above stages are recorded so that an auditable trail is available for the derivation of the control factor. Some materials will move across the whole control factor range from hours to decades during conditioning.

5.4 Calculation of the Radiological Hazard Potential (RHP)

The RHP is calculated from the three factors derived above as follows:

$$\text{RHP} = \frac{\text{Inventory (SITP) x Form Factor (FF)}}{\text{Control Factor (CF)}}$$

Inventory (SITP) = $\sum B_i$ x $(\text{SITP})_i$ Where B_i = number of Bq of radionuclide I, and $(\text{SITP})_i$ = specific toxic potential of radionuclide i

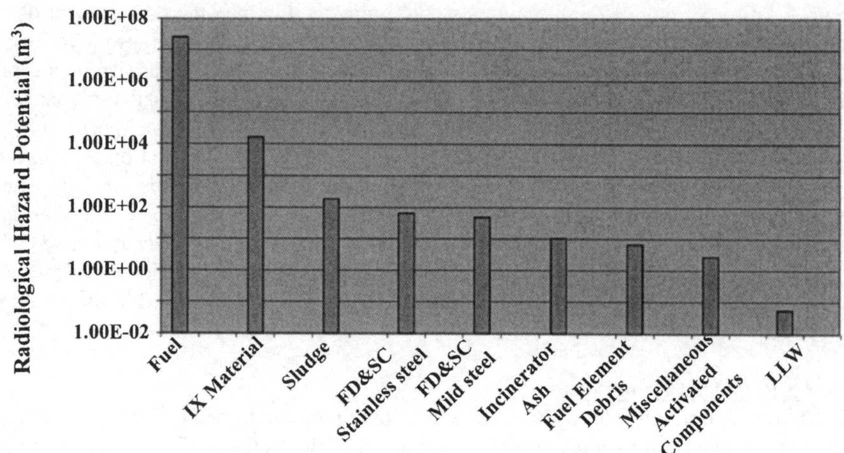

Figure 2 Radiological Hazard Potential – Top impacting Waste Streams

For example, a material of large inventory (SITP) in liquid form (high FF) and with other inherent properties that make it relatively difficult to store (low CF), will have a high RHP value. However, a conditioning programme that retained the Inventory (SITP), converted the material into a monolithic solid (low FF) and reduced the difficulty of storage (high CF), would produce a much lower RHP.

The two modifying factors (FF and CF) can, in the extreme, reduce the value of the RHP for a given inventory by a factor of $10^{-6}/10^{5} = 10^{-11}$. In contrast, the inventory can only reduce through radioactive decay. The above equation leads to an RHP expressed as a volume of water. In practice inventories can range up to 10^{15}, the form factor ranges from 1 to 10^{-6} and the control factor ranges from 1 (hours) to 10^{5} (decades). Hence in numerical terms the RHP has an extremely large range and may need to be represented on a logarithmic scale.

Further detail is available in a number of NDA publications including a recommended list of radionuclides to be used in determining the RHP for UK nuclear materials and the approach to be taken for certain radionuclides where base data is unavailable.[8,9]

6 RADIONUCLIDES

There are currently 2,345 radionuclides identified in the JEF 2.2 Nuclear Data Library.[10] Of these 2,072 are radionuclides with a half life of less than 10 days and are therefore are of no significance to waste management. It might be considered that a 10 day half-life is too short when considering the timescales of decommissioning and waste management. After a maximum of 10 half-lives the activity of a radionuclide would have decreased to about 2% of the starting level and would be considered insignificant. In fact, if the nominal half-life of 10 days is increased to 1-2 years, there is very little change to the number of nuclides eliminated by this criterion.

The remaining 273 potentially relevant radionuclides are subjected to a methodology which considers UK waste streams from commercial reactor fuel and decommissioning waste from final site clearance (SC), in addition to UK safety relevant situations for transport, repository operations and post-closure safety cases. This eliminates 161 radionuclides which are present in insufficient quantities to represent an issue either in terms of operations or for the post-closure repository safety case. The residue of 112 radionuclides are considered relevant to these issues and are the ones that Nirex identify as requiring either an estimate or a measurement of their concentration.

7 RESULTS FOR A MAGNOX REACTOR SITE

Figure 1 shows the total RHP for a two reactor Magnox decommissioning site as a function of time from three months after reactor shutdown. The RHP falls rapidly at first due to removal of fuel from the site and waste retrieval, conditioning and packaging brings about further falls. There is a further gradual reduction in the RHP due to radioactive decay. When all wastes have been conditioned, the only further reduction is again due to radioactive decay until wastes are sent for disposal at a repository. This brings about only a tiny reduction in the RHP which falls further due to decay until final site clearance reduces the RHP to zero. Figure 2 shows the major contributions to the RHP at 3 months after reactor shutdown.

8 LIMITATIONS

There are a number of areas where the RHP and its component factors, have limitations and could result in mis-representation of a hazard or non-discrimination between hazards. For example, nuclear safety cases require the highest reasonably practicable levels of monitoring and intervention to be applied so that risks are reduced to a minimum. If these levels were used for the control factor in the RHP, discrimination would not be possible between the diverse range of stored materials for which the RHP is likely to be applied.

Catastrophic failures, other than those that might arise directly from the properties of the materials (*e.g.* seismic events, aircraft crashes, major vessel failure), should not be considered, as the RHP is concerned with failures which might arise if the material was left unattended and would likely go unnoticed by anyone not specifically tasked with monitoring the facility. Furthermore, specific plant faults should not be considered in this process unless they are directly associated with the identified storage features and could be argued to be aspects of the design proposed by the design engineer. Further limitations could arise from incomplete radionuclide data, either indicating the need for sampling and analysis or a more complete assay for materials no longer available. These materials may contain previously un-quantified elemental trace impurities at the time of manufacture, which have since become activated producing decay products. With NDA encouragement, the industry is developing and maintaining an archive of inactive technological and construction materials, which will enable further sampling and chemical analysis of precursor elements. Any sampling, detection and analysis techniques, which improve the quality of the radionuclide inventory, therefore also serves to improve the reliability of the RHP assessment.

It is possible to aggregate the RHP values for all the waste streams on a site to produce a single value RHP for the site, and this aggregation can continue over several sites although its significance may suffer.

The Chemical Hazard Potential (CHP) addresses chemical hazards associated with a material and can be applied to non-radioactive and radioactive materials.[11] For a material that has both a radiological and a chemical hazard potential (e.g. radioactive asbestos) both of these hazard potentials must be taken into account in deriving the Safety and Environmental Detriment (SED).

Finally, as indicated above it is important that the RHP is only used as one of a number of criteria for assigning priorities in waste management and decommissioning. As mentioned earlier, NDA has recently published a formal prioritisation procedure, which defines the derivation of the RHP together with other parameters (Chemical Hazard Potential (CHP), Facility Descriptor (FD), Waste Uncertainty Descriptor (WUD) and the Ongoing Environmental Detriment (OED)).[3] These five metrics are combined in a simple formula to derive the Safety and Environmental Detriment (SED), which can be applied to a site having multiple facilities. The result is a ranked list of facilities on a given site with those of greatest "threat" having the highest scores. The results for each facility can be summed to give an overall site score.

9 CONCLUSIONS

1. The RHP has evolved into a quantitative measure of the reduction in hazard during waste management and decommissioning on a nuclear site.
2. Use of the RHP is required by NDA, widely accepted by stakeholders and is transparent and easy to use on a routine basis.
3. The RHP addresses the hazard and not the risk of that hazard causing harm.
4. The RHP should be used as one of a number of measures to decide waste management and decommissioning priorities.
5. Improvements to the quality of the radionuclide inventory from developments in sampling, detection and analysis also serves to improve the RHP assessment.

References

1. Command 2919, July 1995, Review of radioactive Waste Management Policy – Final Conclusions, HMSO.
2. Waste Conversion Index, R Jarvis and P Fawcett, RAT 4002, Issue 1, July 2003.
3. NDA Prioritisation Procedure, Document Number EGPR02, Revision 1, July 2006.
4. 2004 United Kingdom Radioactive Waste Inventory – Main Report, DEFRA/RAS/05.002, Nirex Report N/090, October 2005.
5. Publication 68, Dose Coefficients for Intakes of Radionuclides by Workers, Group of Committee 2 of International Commission on Radiological Protection, 1995.
6. ICRP Publication 72, Age-dependent Doses to the Members of the Public from Intake of Radionuclides: Part 5, 1995 Compilation of Ingestion and Inhalation Coefficients, Annals of International Commission on Radiological Protection.
7. National Radiological Protection Board (NRPB) W41, Generalised Habit Data for Radiological Assessments, K R Smith and A L Jones, May 2003.
8. Instruction for the Calculation of the Radiological Hazard Potential, Nuclear Decommissioning Authority, 13 July 2006, Document No. EGPR02-W101, Rev. 2.
9. Radioisotopes in the Radiological Hazard Potential, Nuclear Decommissioning Authority, Document Number EGR004, June 2006.
10. The Joint Evaluated File (JEF-2.2) Nuclear Data Library, April 2000, JEFF Report 17, OECD NEA.
11. Instruction for the Calculation of the Chemical Hazard Potential, Nuclear Decommissioning Authority, 13 June 2006, Document No. EGPR02, Revision 1.

Acknowledgment - The authors gratefully acknowledge support and permission to publish from the NDA and Magnox Electric Limited.

DETERMINATION OF TRITIUM RADIONUCLIDE AND LITHIUM PRECURSOR IN MAGNOX REACTOR STEELS

W.A. Westall, R.E. Streatfield and D.J. Hebditch

Engineering & Technical Services, Magnox Electric Limited, Berkeley Centre, Berkeley, GL13 9PB, U.K.

1 INTRODUCTION

Tritium, ^{3}H, is produced in nuclear reactors from the various neutron activation reactions of ^{2}H, ^{6}Li and ^{10}B impurities in structural materials.

Deuterium is in very low concentration. Lithium has an atomic weight of 6.94 and the abundance of ^{6}Li is around 7% in natural Li. The main reaction product of ^{10}B is ^{7}Li which does not generate ^{3}H but there are other, minor reactions that do. Except in boron steels, the activation of ^{6}Li predominates. Another source of ^{3}H in fission reactors is the low yield, ternary fission of fuel (~130 x 10^{-6} atoms per fission product pair). In Magnox gas-cooled reactors, ^{3}H from ternary fission is mainly retained in the metallic uranium fuel and its cladding but some is released into the coolant circuits, where it may possibly diffuse into structures within the primary vessel. Tritium is a low energy β emitting radionuclide of low radio-toxicity and with a half life of 12.3 years.

In calculations of the radioactive inventory of the activated steel structure of Magnox reactors and of routinely removable components such as neutron absorber bars, ^{3}H has generally received little attention. This is because it is not a major radiological hazard when compared to the gamma dose rate of other activated species and the concentration of its precursor, ^{6}Li, was commonly assumed not to be significant. Now that the miscellaneous activated component (MAC) storage vaults of Magnox nuclear power stations are being emptied as part of the decommissioning process and their contents are being packaged for final disposal, the radioactive inventories for all radionuclides potentially present must be considered.

A fit-for-purpose estimate of the individual activities is needed for waste disposal purposes and calculation of neutron activation of precursors is often the simplest method, where contamination is unimportant. For ^{3}H, the concentration of the ^{6}Li precursor is required. The mobile nature of ^{3}H is a complicating factor since, in principle, it may remain *in situ* or diffuse within the bulk material. The aim of this work was to provide some answers to the above questions by determining the concentrations of Li in reactor steels and to compare the predicted levels of ^{3}H with values measured in reactor surveillance specimens. Preliminary attempts to measure Li in reactor steels by ICP-OES (inductively coupled plasma - optical emission spectroscopy) were not successful and yielded concentrations over the range 4 to 45 $\mu g\ g^{-1}$; far higher than expected. Results of measurements by a variety of other techniques, including ICP-MS (inductively coupled plasma – mass spectroscopy), SIMS (secondary ion mass spectroscopy), SEM-EDX (scanning electron microscope – energy dispersive X-ray), and NAA (neutron activation analysis) with subsequent radiochemical analysis are given below.

2 METHODS AND RESULTS

2.1 Measurement of Lithium in RPV Steel by ICP-MS

Eight samples of un-irradiated reactor pressure vessel (RPV) steel, two each from Trawsfynydd (TRA), Dungeness A (DNA), Sizewell A (SXA) and Bradwell (BWA) reactors were analysed. ICP-MS analysis was carried out using a high resolution magnetic sector instrument.[1] Despite the sensitivity of this method, i.e. lower limit of detection (LLD) of around 8 pg g^{-1} for procedural blanks, it failed to detect Li and achieved a detection limit of 80 ng g^{-1}, which was well above the level of interest. However, the results were consistent and did show that the Li concentration was well below that found from the earlier analytical attempts (ICP-OES) and below the levels conservatively assumed in the waste inventory assessments.

The reason for the inadequate detection limit was that ICP-MS was intolerant of high levels of dissolved solids especially a matrix composed largely of a single element, Fe in this case. Therefore very dilute solutions needed to be used, e.g. 10,000 fold dilution, to avoid suppression effects. Even though the initial detection limits in solution were pg g^{-1}, they increased to the levels reported after the required dilution. Possible methods of improving the detection limits include removal of the Li from the Fe matrix by chemical separation following sample dissolution thereby allowing a much lower dilution factor or, less desirably, using the procedure of standard additions rather than external calibration.

2.2 Measurement of Lithium in Steel by SIMS and SEM-EDX

As ICP-MS as applied to the samples was not sensitive enough to detect Li directly, alternative methods were sought. SIMS was demonstrated to be adequately sensitive using two samples of DNA RPV steel and so further experiments were carried out to identify the environment of the Li and its concentration. The work was carried out in three stages.[2] The first stage was to test the feasibility of the approach, secondly to characterise the micro-structural environment of the Li and finally to quantify the concentration of Li in steel.

SIMS is a technique frequently used in the semiconductor industry for ultra-trace chemical analysis and therefore could easily achieve the required sensitivity. The material to be analysed is bombarded with a beam of ions which generates secondary ions from the matrix which are ejected from the surface and analysed by the mass spectrometer. The equipment used was a Cameca 3F with an 8.5keV O_2^+ primary ion beam. The beam current was in the 1-5 μA range forming a raster across a 250 μm field-of-view, out of which the inner 150 μm diameter area was analysed. An example of the results of the experiments is shown in Figure 1 (these are actually the quantitative SIMS profiles obtained after calibration). The level of the Li varies with position, (proportional to bombardment time and reaching a depth of 100 μm), indicating that most of the Li is not homogeneously distributed within the matrix but possibly at grain boundaries as inclusions.

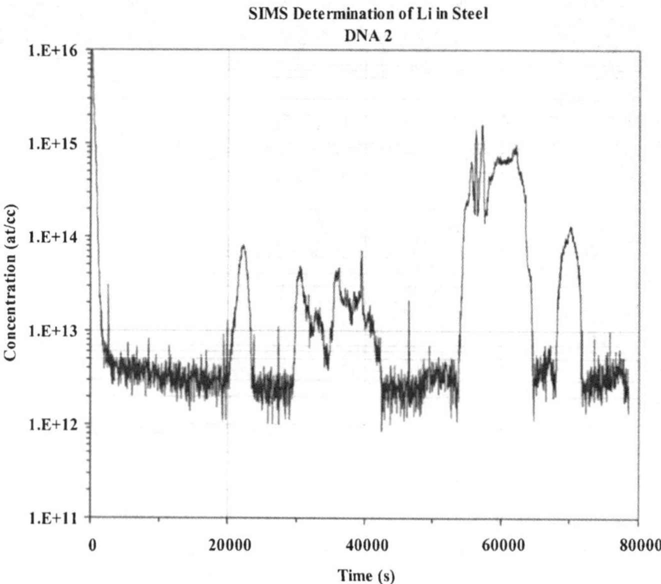

Figure 1. *Lithium Profile From SIMS Analysis of Dungeness A RPV Steel Sample 2*

Further investigation of the Li environment was performed using SEM-EDX analysis where the inclusion could be imaged using SEM and the chemical environment characterized by EDX. Analysis of the SIMS crater base was carried out using a Jeol 6100 microscope at 20 keV using an Oxford ISIS detector with ultra-thin window. Figure 2 shows various 250 µm size SIMS image fields. When a maximum in the Li signal was located, the SIMS was switched to imaging mode and Li, Fe, carbon (C) and aluminium (Al) images of the crater base taken. A definite Li-containing spot is seen and found to be associated with dominant Al. The Li signal investigated was discovered to be associated with a micron sized Al inclusion which also included the additional species sodium (Na), sulphur (S), boron (B) and calcium (Ca) with depleted C. The Al mass number 27 (mass to charge ratio – m/z 27) was so intense that imaging was carried out using the oxide signal at m/z 43.

The final stage of the SIMS analysis was to quantify the profiles obtained in the initial trial. For this a standard was prepared by implanting a known dose (atoms cm^{-2}) of $^{7}Li^{+}$ ions into the target. This was analysed by SIMS, tracking both $^{7}Li^{+}$ and matrix ion species, and the crater depth measured by profilometry. This allows calculation of a relative sensitivity factor (RSF) which converts the ^{7}Li-to-matrix ion intensity ratio into a Li concentration in atoms cm^{-3} as a function of depth. When the unknown was measured, with ^{7}Li and the same matrix ion signals recorded, then the RSF was applied to the signal ratio to give the Li level in the unknown.

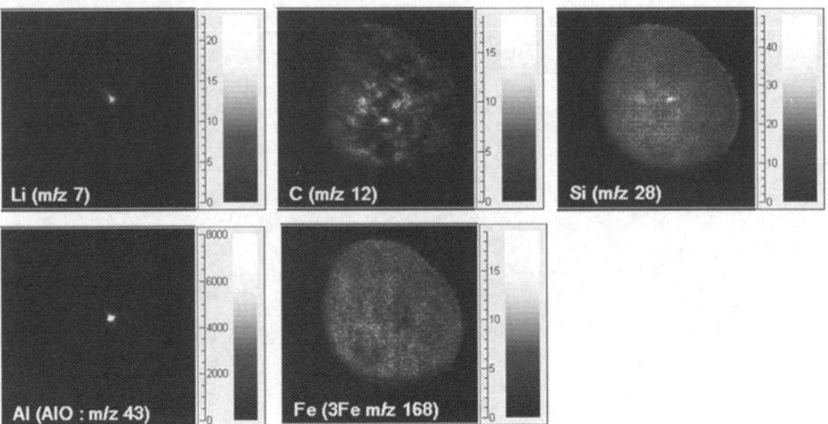

Figure 2. *SIMS Images from DNA2 steel sample crater base*

The quantitative data from a profile is given in Figure 1. To arrive at a Li concentration the number of atoms per cm^3 for each of the peaks was averaged to give an overall value. This was then divided by the number of atoms per cm^3 of the iron matrix to give the ratio of Li atoms to iron atoms and converted to ng g^{-1} by multiplying by the ratio of the atomic weights of Li and Fe. The measured Li concentrations, for samples DNA1 and DNA2 were respectively: 0.01 ng g^{-1} and 0.11 ng g^{-1} giving an average of 0.06 ng g^{-1}.

Uncertainties in the above values come from the dependence of the number and sizes of the inclusions in the matrix and non-linearities in the dependence of the Li ion yield with concentration. The main contributor to the large disparity between the Li concentrations found for samples DNA 1 and 2 is regarded as the local variation in the number and size of the inclusions. To reduce the uncertainty in the results a larger number of inclusions over a wider area would need to be sampled but even then the natural variation in the microstructure may still give a difference between the samples for practicable numbers of measurements. The calibration was for Li on a Fe substrate whereas for actual samples the Li was associated with Al inclusions. This may result in a different RSF than used in the quantification of the Li to matrix ion intensity and so a variation from the reported Li concentration. The combined effect of the sources of uncertainty are compatible with the ten-fold difference between the measured results so that the present data are currently regarded as contributing to a best estimate of magnitude.

2.3 Measurement of Tritium in Irradiated Specimens

Six samples of irradiated TRA RPV steel were obtained from the material archives at Berkeley Centre. The samples were from RPV surveillance specimens, which were placed within the reactor at known positions and withdrawn periodically during the operational life of the reactor to monitor the effects of radiation damage on the RPV. The samples used here were from Charpy specimens, which had been in the reactor for its entire operational life of 26 years. Samples of approximately 1 g were analysed for ^{3}H and ^{14}C by at the Springfields Laboratory using the standard technique of heating the specimen to a high temperature under a stream of oxygen.[3] The tritiated water produced was collected in a water trap and the ^{14}C bearing carbon dioxide was collected in a trap containing Carbosorb. The collected samples were analysed using liquid scintillation counting. The results are presented in Table 1, where data are corrected for radioactive decay to 14th December 2004 and uncertainties are given at the 2σ level based on counting statistics only.

From the measured values of ^{3}H and knowing the activating neutron flux and irradiation time the concentration of the precursor can be found. The irradiation time has been taken as 26 years and the flux has been obtained from the Magnox Pressure Vessel Dosimetry Manual: Issue 9. The fluxes are given in displacements per atom (dpa), which may be converted to reaction rates for Li by multiplying by the ratio of the neutron capture cross-sections of ^{6}Li to Fe dpa in barns per atom and dividing by the irradiation time. It has been assumed that the reaction is entirely due to thermal neutrons due to the ^{6}Li resonance integral being less than half that of the thermal cross-section. The measured ^{3}H was corrected for radioactive decay both during and after irradiation. The build-up to saturation during activation has been accounted for by a factor of $1-\exp(-\lambda t)$ in the activity, where λ is the decay constant for radioactive decay and t the period of irradiation. This equates to 0.769 over the 26 years activation. The activity was also decayed to a common date. The Li was calculated as follows:

$$^{6}\text{Li concentration (atoms g}^{-1}\text{)} = \frac{^{3}\text{H (Bq g}^{-1}\text{)}}{\text{dpa}} \cdot \frac{\sigma_{Fe}}{\sigma_{Li}} \cdot t_a$$

Where:

dpa is the number of displacements per atom for iron
σ_{Fe} Cross section, barns (at 2200 m s^{-1} neutron velocity) for Fe
σ_{Li} Cross section, barns (at 2200 m s^{-1} neutron velocity) for ^{6}Li
t_a is the activation time (s)

The ^{6}Li concentration in atoms per g was converted to ng g^{-1} natural Li value and the results are presented in Table 2. The six samples were from three canisters, all of which experienced predominantly thermal fluxes. Two of the canisters (containing samples 30006, 30010, 35031 and 35038) were from below the core with an epithermal flux value twice that of the other canister (samples 30114 and 30115) which was positioned at the side of the core. The calculated Li concentrations are generally consistent for all samples apart from specimen 35031 where the ^{3}H result was affected by contamination with ^{14}C. The average value for the predicted natural Li concentration in RPV steel, excluding the result for specimen 35031, was 0.3 ± 0.1 ng g^{-1}.

Table 1 : *Measured Values of ^{3}H and ^{14}C in TRA RPV Surveillance Specimens*

Surveillance Specimen ID	Power Station Sample ID	H-3 (Bq g^{-1})	C-14 (Bq g^{-1})
30115	WAW 75/1/0	11.6 ± 0.16	234 ± 1.4
30006	WAW 75/2/0	1.03 ± 0.077	28.6 ± 0.41
35038	WAW 75/3/0	1.90 ± 0.22	32.1 ± 3.3
35031	WAW 75/4/0	8.63 ± 1.0	43.8 ± 3.4
30114	WAW 75/5/0	8.86 ± 0.14	256 ± 1.6
30010	WAW 75/6/0	0.992 ± 0.076	28.1 ± 0.40

Table 2 : *Calculated Values of Lithium in TRA Surveillance Specimens*

Specimen ID #	Thermal flux dpa	Uncertainty %	Tritium Activity Bq/g	Predicted Natural Li ng g^{-1}
30006	12.2 x 10^{-6}	55	1.03 ± 0.08	0.30 ± 0.17
30010	12.2 x 10^{-6}	55	0.99 ± 0.08	0.29 ± 0.16
30114	269 x 10^{-6}	31	8.86 ± 0.14	0.12 ± 0.04
30115	269 x 10^{-6}	31	11.60 ± 0.16	0.15 ± 0.05
35031	13.3 x 10^{-6}	55	8.63 ± 1.00	2.30 ± 1.29
35038	13.3 x 10^{-6}	55	1.90 ± 0.22	0.51 ± 0.28

2.4 Measurement of ^{6}Li in Steel by NAA and Radiochemical Analysis of ^{3}H

Another approach to the determination of Li in RPV steel was by neutron activation analysis (NAA) followed by radiochemical analysis for ^{3}H. This was far more complex than the ICP-MS analysis as it involved the irradiation of the inactive steel specimens in the CONSORT reactor at Ascot and transport of the activated steel to the Springfields Laboratory for radiochemical analysis.[3] Six inactive specimens of RPV steel from TRA were irradiated for 63 hours in a neutron flux comprising flux values of: 960 x 10^9 n cm^{-2} s^1, 44 x 10^9 n cm^{-2} s^{-1} and 300 x 10^9 n cm^{-2} s^{-1} for thermal, epithermal and fast regions of the energy spectrum respectively.

The samples were transported for analysis immediately after irradiation. Prior to the irradiation of the assay samples, a trial irradiation had been carried out to test the feasibility of this approach which identified a problem in the radiochemical analysis. An induction furnace was again used to melt the steel samples under a stream of pure oxygen gas. ^{35}S is another activation product of steel but with its relatively short half life of 87.2 days is not generally of consequence to decommissioning measurements. However, due to the very short time between irradiation and analysis the ^{35}S was still present in the specimens. ^{35}S is volatile and was driven off (as mixed oxides of sulphur) and collected with the ^{3}H^{1}HO and possibly $^{14}CO_2$ under the analysis conditions and interfered with the liquid scintillation counting for both. Therefore, H_2O_2 was added to the ^{3}H trap to oxidise sulphur compounds to sulphate followed by sample distillation to remove the ^{35}S from the counting of the ^{3}H and ^{14}C. Results were obtained for three of the six samples and are presented in Table 3. The Li concentration data was calculated in the same way as for the Magnox RPV radioactive inventory assessments and are presented in Table 3, where data are corrected for radioactive decay to the 14th December 2004 and uncertainties are given at the 2σ level based on counting statistics only. Two of the three values lay close together but the third was significantly greater and the average value for the three Li concentrations was found to be 0.5 ± 0.3 ng g^{-1}.

Table 3 : *Tritium in TRA RPV Steel measured by NAA and Radiochemical Analysis*

Sample ID	^{3}H Bq g^{-1}	Derived Lithium Concentration ng g^{-1}
WRL/0382	1.33 ± 0.22	0.40 ± 0.07
WRL/0384	1.00 ± 0.26	0.30 ± 0.08
WRL/0387	3.00 ± 0.25	0.90 ± 0.08

3 DISCUSSION

Work on the waste management of retrieved MAC at TRA has found that even very low concentrations of Li could give rise to quantities of ^{3}H which require assessment for disposal in any future Intermediate Level Waste (ILW) repository. NIREX provides guidance in the form of activity concentrations above which radionuclides are considered to be present in quantities that are significant with respect to transport, operational and post closure safety.[4] Such concentrations are known as Guidance Quantities (GQ) and determine the type of information required for radionuclide data recording purposes. When a radionuclide is present at greater than its GQ, then it is likely that the radionuclide will require detailed determination at the individual waste package level. If the radionuclide concentration is below its GQ then details of the simple bounding concentrations used in the significance test should be adequate. Therefore an improved understanding of the levels of ^{3}H in activated steels will be important in reducing uncertainty in the costs of waste management of irradiated metal alloys. It was estimated in the TRA MAC study that if the Li concentration was > 5 ng g^{-1} the ^{3}H produced could, depending on the circumstances, exceed the NIREX Guidance Quantity (GQ).[4]

The intention of this work was to measure the concentration of Li in examples of un-irradiated reactor steel from which the likely ^{3}H activity concentration could be calculated using suitable values for the neutron flux.[5] Calculated values could then be compared with the ^{3}H activities measured in examples of steel irradiated in a Magnox reactor. From this it was hoped to ascertain whether, given the concentration of Li, the ^{3}H activity could be reliably calculated or whether the effects of ^{3}H formation from ternary fission and subsequent diffusion would prevent this. Whilst the ^{3}H activity in activated steel was measured in a straight forward manner, the measurement of Li in steel was difficult. The conventional analytical techniques of ICP-OES and ICP-MS respectively failed to achieve an adequate detection limit, so alternative techniques were tried. SIMS was sensitive enough to detect Li but is a microscopic technique and so was prone to large uncertainties when used to predict the bulk concentration. NAA followed by radiochemical determination allowed the determination of ^{3}H but initially suffered interference from another activated radionuclide (^{35}S).

SIMS was sensitive enough to easily detect Li at the low levels encountered but suffered some problems when quantification was attempted. SIMS is essentially a microscopic technique but in this case it was applied to determining the bulk concentration with some inherent uncertainties. The Li was found to be associated with Al inclusions and so if the Li quantity is proportional to that of the Al and there is a range of Al inclusion sizes, then a large number of inclusions must be sampled to reduce the sampling uncertainty. This was not possible in the current project which has given rise to a potentially large, up to a factor of 5, sampling uncertainty in the concentration value. Aluminium has been measured in Magnox RPV steels and is found in concentrations over the range 20 – 140 μg g^{-1}. Therefore if the Li concentration is dependent upon that of the aluminium a variation in concentration of approximately an order of magnitude could be expected.

The RSF was determined using a Li doped Fe target but the Li is localised in reactor steels and strongly associated with Al. It is considered that the true RSF may vary from that determined from the Fe target by a factor of three, which gives an overall uncertainty of approximately a factor of 6.[4] The average value obtained from SIMS, 0.06 ± 0.05 ng g^{-1} should therefore be used as an indication of the likely range of Li concentration after consideration is given to the additional uncertainty.

The radiochemical analysis for ^{3}H of RPV surveillance specimens, neutron activated in reactor, was straight forward and allowed the effective precursor Li concentration to be calculated by standard methods. However, the activating neutron fluxes had relatively high associated uncertainties leading to similar uncertainties in the Li concentration. The calculated Li concentration depends upon the measured ^{3}H activity. So that if ^{3}H has diffused out of or into the specimen over its time in the reactor or subsequent storage time outside of the reactor, then the predicted Li concentration will have been underestimated or overestimated accordingly. The values for the Li concentrations obtained were consistent apart from one value which was thought to be due to contamination of the tritium trap with ^{14}C. The average value for the Li concentration, ignoring the spurious result, was 0.3 ± 0.1 ng g^{-1}.

NAA immediately followed by radiochemical analysis reduced the possibility of any ^{3}H formed diffusing out of the specimens. The activating neutron flux was well characterized with an uncertainty of approximately 6% and so the dominating uncertainty will be from the radiochemical analysis which varied from 8% to 25%. There was a complicating factor where ^{35}S formed during the neutron irradiation had to be removed before the ^{3}H analysis, otherwise it would completely obscure the analytical signal for tritium. It is believed that the removal of ^{35}S was completely successful so that the results obtained are reliable. The three results obtained were not as self consistent as for the RPV surveillance specimens with one value being more than twice that of the others. There is no reason to reject the high value and the average Li concentration for all three values is therefore 0.5 ± 0.3 ng g^{-1}.

Comparing the tritium-derived, predicted concentrations for Li from analysis of surveillance specimens and from NAA, there is no significant difference between the average values. This suggests there is no significant diffusion of ^{3}H either out of or into the steel during the time in the reactor or in storage. Accordingly, the ^{3}H activity can be reliably calculated from the Li concentration in reactor steels. From the results obtained here, the average Li concentration calculated from the values obtained from the analysis of the surveillance specimens and NAA and radiochemistry on RPV steel is 0.4 ± 0.2 ng g^{-1}. At the 95% confidence level, the upper limit value is 0.8 ng g^{-1}.

Measurements have only been made on a few RPV steel specimens and although an average value for the Li concentration has been derived there is still a large associated uncertainty. Analysis of trace impurities in reactor materials has identified that the concentrations follow a log-normal distribution and therefore it is possible to have a large range of possible concentrations of a trace impurity in any given material. It has not been possible with the current set of data to demonstrate that Li behaves in a similar manner. For this more samples need to be analysed to give greater confidence in the results. Further samples could be analysed by NAA followed by radiochemistry but this is very expensive and has relatively high associated uncertainties. For SIMS to be used with confidence further development work will be necessary to investigate the RSF for the true environment of the Li and extended analyses would need to be carried out to sample larger numbers of inclusions. An alternative is to develop a sample clean-up method for use with ICP-MS. Once a method is developed, it should be cheaper and easier to apply to larger numbers of samples thereby increasing the confidence in the analytical results. This method could be applied to a variety of materials that are stored in MAC vaults such as stainless steel, zirconium and nimonic alloys, for which there are currently no data.

Table 4 *: Summary of Measured and Calculated Lithium and Tritium Data*

Experimental Technique	Lithium level ng g^{-1}	Tritium activity Bq cm^{-3}	Mean Li level Measured/Calculated
ICP-OES Li Measurement	4,000 – 45,000 *Measured*	Not present	Excessive Li values
ICP-MS Li Measurement	< 80 *Measured*	Not present	Poor LLD for Li
SIMS and SEM-EDX Li Measurement	0.01 and 0.1 *Measured*	Not present	Mean Li = 0.06 ng g^{-1} Tenfold uncertainty
Magnox reactor Radiochemical ^3H	0.3, 0.29, 0.12 0.15, (2.30), 0.51 *Calculated*	1.03, 0.99, 8.86, 11.60, (8.63), 1.90 *Measured*	Mean Li = 0.3 ng g^{-1} ± 0.1 ng g^{-1}
CONSORT reactor Radiochemical ^3H	0.40, 0.30, 0.90 *Calculated*	1.33, 1.00, 3.00 *Measured*	Mean Li = 0.5 ng g^{-1}± 0.3 ng g^{-1}

In summary, four experimental/calculational techniques, ICP-OES, ICP-MS, SIMS-SEM-EDX and NAA with radiochemical analysis, were used to determine the concentration of Li in reactor steels either by direct measurement or by calculation from measured ^3H values. Li measurements from the first two techniques, ICP-OES, and ICP-MS, were discarded due to excessive values/excessive limit of detection. SIMS/SEM-EDX measurement of Li and two approaches to neutron activation of steels samples with radiochemical measurement of ^3H values used to calculate Li concentrations yielded 10 valid data points spread over one order of magnitude, see Table 4. In particular, the two neutron activation approaches using differing flux, time and temperature conditions yielded good consistency of calculated values of Li. Consequently, the average Li concentration in Magnox RPV steel is considered to be < 1 ng g^{-1} and estimated as 0.4 ± 0.2 ng g^{-1} x (1Б).

There is still a level of uncertainty associated with the data but from the measurements the Li concentration can be said to be lower than 1 ng g^{-1}. This is an improvement as previously relatively high Li concentrations had been assumed leading to correspondingly large calculated values for the ^3H activity in activated reactor steels. At present this is important for items routinely removed from reactors during operation such as absorber bars which are to be removed from their storage vaults for repackaging and subsequent disposal. It should be possible to estimate the ^3H activity without having to sample and analyse these often highly active items.

4 CONCLUSIONS

The concentration of natural lithium (tritium-precursor) in un-irradiated Magnox RPV steel has been estimated directly using SIMS-MS and SEM-EDX and indirectly from measuring tritium induced by neutron activation of mainly ^6Li. Tritium was measured in surveillance specimens irradiated during the 26 year operational life of a Magnox reactor followed by radioactive decay of approximately 15 years. For comparison purposes, inactive archive RPV steel was also irradiated in the CONSORT reactor followed immediately by radioanalysis for tritium. In this way, the possibility of diffusive transfer of tritium into or out of the steel during residence in the reactor could be evaluated.

1. Based on various approaches, analytical techniques, calculations and available samples, the average lithium concentration in Magnox RPV steel is considered to be < 1 ng g^{-1} and estimated as 0.4 ± 0.2 ng g^{-1} (1Б)

2. Establishing an average Li concentration < 1 ng g^{-1} should enable estimation of ^{3}H activity for waste management purposes without the need to sample and analyse high dose rate wastes, e.g. neutron absorber bars.

3. Lithium appears to be associated mainly with aluminium inclusions within the steel.

4. Given the dependence on trace impurity levels and their distribution, further analytical data are needed to increase the confidence in the estimated value of the average concentration of natural lithium in Magnox RPV steel.

5. The development of an improved analytical method for measurement of Li in RPV steels is desirable and the use of ICP-MS with chemical removal of matrix material, Fe, from dissolved samples may be optimal.

6. Similar steel samples irradiated under differing flux, time and temperature conditions yielded good consistency of calculated values of Li precursor.

7. Preliminary indications are that tritium activity may be predicted from the natural lithium concentration alone and no evidence was observed of significant tritium diffusion into or out of steel activated in a Magnox reactor.

References

1 K.E. Jarvis, Determination of Lithium in Steel Samples by High Resolution Magnetic Sector ICP-MS, 7 December 2004, Viridian Partnership Report, for British Nuclear Group.
2 A.J. Pidduck, M.R. Houlton and G.M. Williams, Investigation into the Form of Li present in RPV Steel Sample DNA2, 3 February 2005, QinetiQ Report, Malvern, for British Nuclear Group.
3 B. Jackson, "Decommissioning Studies – Analysis for H-3 and C-14 in Trawsfynydd RPV Samples and Imperial College Irradiated Steel Specimens", March 2005, BNFL Engineering Advice Note, NSTS/GEN/EAN/0067/05.
4 Customer Guidance on the Requirements for Waste Package Radionuclide Inventories, NIREX Technical Note Prepared by J. Jowett (SERCO), Document Number 384094, December 2001, Report Number SA/R/PSEG/04441, Draft 3.
5 W.A. Westall, Determination of Tritium and Lithium in Reactor Steels, British Nuclear Group, Berkeley Centre, April 2005, Report No. E&T/REP/GEN/1627/05.

Acknowledgment - The authors gratefully acknowledge support and permission to publish from the NDA and Magnox Electric Limited.

SEQUENTIAL DETERMINATION OF Ca-41/45 AND Sr-90 IN AN ACTIVATED CONCRETE CORE.

F. Rowlands*, P. Warwick and I Croudace

GAU-Radioanalytical, National Oceanography Centre, Southampton, SO14 3ZH, U.K

1 INTRODUCTION

Bioshield concretes are subject to a large integrated neutron flux over an extended period of time, leading to neutron activation of the constituents of the concrete. Typical neutron activation products having half lives up to 100 years include ^{60}Co, ^{152}Eu, ^{55}Fe and ^{63}Ni. Also produced are ^{45}Ca (t½ = 164 days) and ^{41}Ca (t½ = 103000 years). Owing to the high Ca content of concretes, this means that the long-lived ^{41}Ca in bioshield concrete could be a potential problem for disposal because of the large volume of material (although of low radiotoxicity). ^{90}Sr is a fission product, and as such should not be present in concrete which has undergone neutron activation. If the concrete contains significant concentrations of ^{235}U, however, ^{90}Sr could be produced through nuclear fission reactions (Hou *et al.*, 2005).

Radiochemical methods which allow sequential determination of radionuclides are often desirable for reasons of speed and cost-effectiveness, for example in mixed waste streams or in waste characterization for nuclear decommissioning. Although a variety of methods are in use for the analysis of $^{41/45}$Ca in concretes, they tend to use large reagent volumes and involve many separation stages. This is due to the fact that the Ca concentration in concretes is high (>40 000 ppm), and other alkaline earth metals have to be removed sequentially from the samples, rather than removing the Ca. Many methods using cation exchange resins are listed (e.g. Korkisch, 1989), but they require less than 0.1% Ca, and do not remove the other alkaline earth metals as the final Ca determination is mass spectrometric rather than radiometric.

The ideal chemical separation method should be relatively simple, have few separation stages and should be time / cost effective. Itoh *et al.* (2002) used a combined anion and cation exchange column method for the isolation of Ca from other alkaline earth metals. This 3 column separation involved 6 chemical separation techniques following two $HF/HClO_4$ digests. The decontamination factors for ^{152}Eu and ^{60}Co are quoted as $>10^3$. The final limit of detection for ^{41}Ca for this method was 8 Bq/g, based on a measurement by X-ray spectrometric determination.

Other methods currently in use for ^{41}Ca determination in concretes are based entirely on precipitations for purification of the Ca. Suarez *et al.* (2000) utilise a chromate precipitation for removal of Ba and Ra from the solution. This is a pH dependant precipitation, and the chromate solution resulting from this requires a separate waste stream. The method has a total of 9 steps after sample dissolution, and has decontamination factors of $>10^4$ for Ba and Co, $>10^5$ for Sr and $>10^3$ for Eu. The minimum detectable activity by liquid scintillation counting is 0.29 Bq/g for ^{41}Ca (0.034 Bq/g ^{45}Ca).

The method described by Hou *et al.* (2005) utilizes a series of pH or concentration controlled precipitations for the purification of Ca, with 7 method stages, including the chemical recovery determination. The determination of ^{90}Sr, however, involves a further 3 chemical separation steps, and a further 3 weeks for a result. Decontamination factors are $> 10^5$ for Eu, Cs, Sr, Fe, Ni, Co.

The method described in this paper utilizes both precipitations and ion extraction chromatography for purification of Ca and Sr, with MnO_2 resin replacing the pH controlled Ba-chromate precipitation required for removal of Ba and Ra from the solution. It is a relatively fast, easy method, and as such is cost and time effective which are important considerations for a commercial laboratory.

The MnO_2 resin (Eichrom) used in this method has a very high uptake of Ba and Ra from neutral solutions (figure 1a), and minimal retention of Ca. The majority of Ca loaded onto the MnO_2 column is eluted using 5 ml of column wash solution (figure 1b). These two factors combine to make the MnO_2 resin desirable for the Ca extraction method.

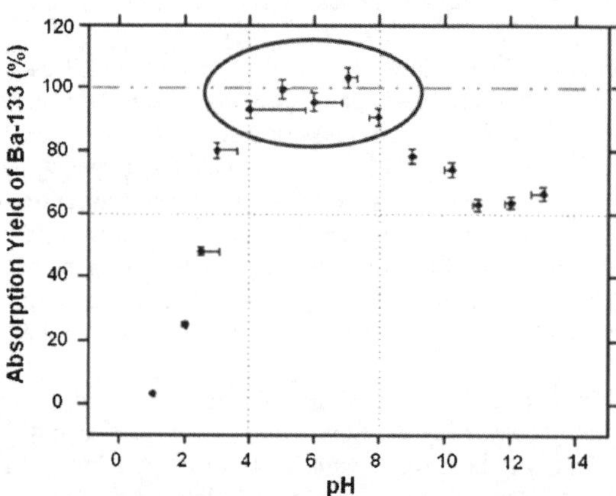

Figure 1a. *Adsorption of Ba-133 onto MnO_2 resin with changing pH (Bombard, 2005).*

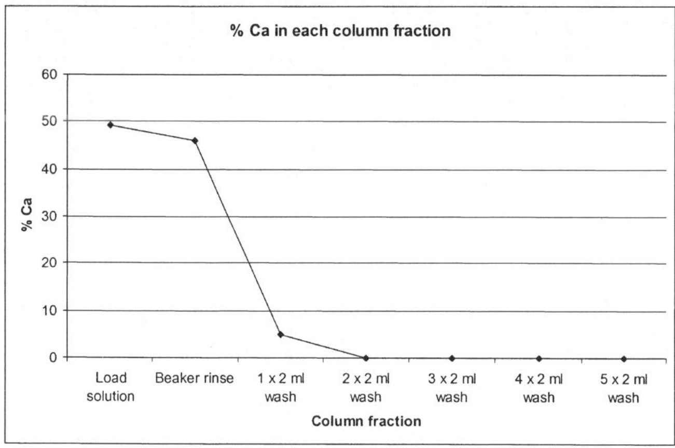

Figure 1b. *Elution of Ca from MnO2 resin at pH 7.*

2 MATERIALS AND METHODS

Reagents:

Eichrom Sr resin
Eichrom MnO_2 resin
4 % ammonium oxalate solution
Saturated KIO_4 solution
10 mg/ml Fe carrier solution
Radionuclide standard solutions of Ca-41, Ca-45, Sr-85 and Sr-90

Equipment:

Canberra well-type HPGe gamma detector for determination of ^{85}Sr
Liquid scintillation counter (Wallac 1220) for determination of $^{41/45}$Ca by liquid scintillation counting and ^{90}Sr by Cerenkov counting (counting of ^{90}Y daughter).
X-ray fluorescence spectrometer (Panalytical Magix-Pro XRFS) for stable Ca determination.

Radiochemical separation method (see figure 2):

- An aliquot of the ground sample is digested using two aqua regia attacks.
- Ca is removed from the resultant solution by Ca-oxalate precipitation, and the precipitation is ashed at 450 °C overnight.
- The residue is dissolved in aqua regia, Fe carrier is added and an $Fe(OH)_3$ precipitation carried out.
- Saturated Na_2CO_3 is added to the supernatant, to form a $CaCO_3$ precipitate.
- This precipitate is dissolved and loaded onto a Sr resin column. The date and time of Y removal from the column is recorded. All the column eluents are collected for further Ca purification, and the Sr fraction is eluted using water, and counted by Cerenkov

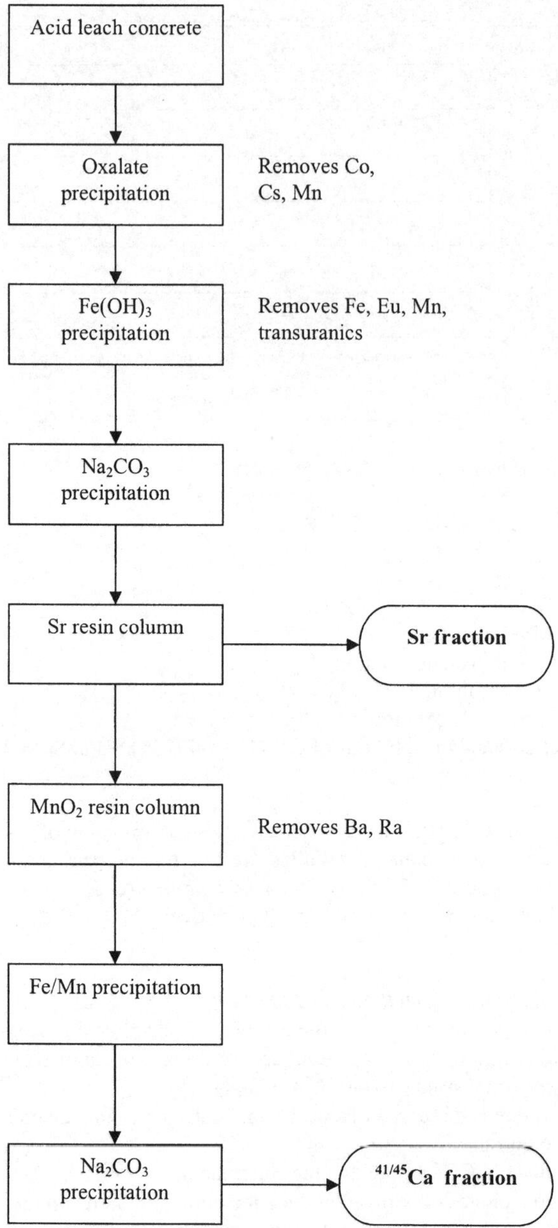

Figure 2. *Schematic of the Ca separation method.*

counting 3 times over 10 days to determine the ^{90}Y in-growth. The Ca fraction from the Sr column is evaporated to dryness.

- The Ca residue is dissolved in neutralised acetic acid and loaded onto the MnO_2 resin column. All MnO_2 resin load and wash solutions are collected and evaporated to dryness.
- Residue is dissolved in HCl, Fe carrier and KIO_4 are added, and a Fe/Mn co-precipitation is carried out.
- Saturated Na_2CO_3 is added to the supernatant to form the final $CaCO_3$ precipitate. This is dried and weighed prior to dissolution and liquid scintillation counting.

2 RESULTS AND DISCUSSION

The method described above was tested on concretes containing ^{41}Ca, concretes spiked with ^{45}Ca and ^{90}Sr, and blank concretes. A sample spiked with Amersham QCYK mixed gamma solution was also run to determine what radionuclides were removed with which stage of the method. The results (Figure 3) show that the final $^{41}CaCO_3$ fraction was free of any of the added gamma emitting radionuclides. Examination of the LSC spectrum for the Ca-41 sample only showed a signal in channels 1-125 (the measurement window).

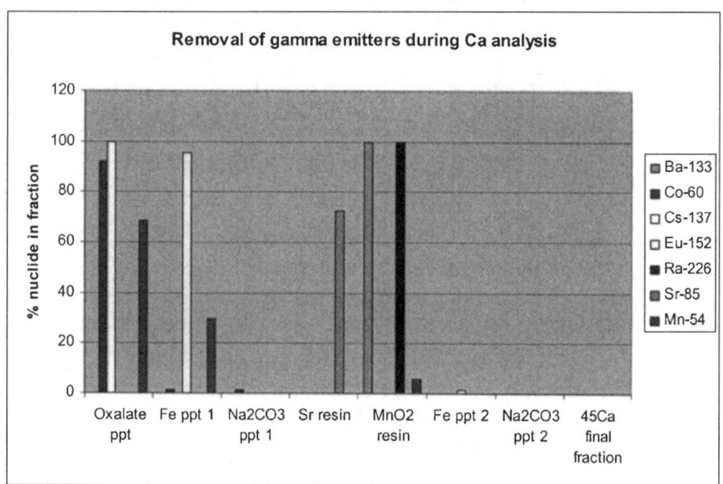

Figure 3. *Removal of other radionuclides during Ca purification.*

The method decontamination factors were determined using additions of Fe, Ba, Cs, Eu, Co and Ni (all commonly found in activated concrete samples). The final fraction was analysed by ICP-MS and ICP-AES, showing that the decontamination factors for Fe and Ni were $> 10^4$, and for Co, Cs and Eu were $> 10^6$. The decontamination factor for Ba was only ~ 7, implying that the MnO_2 resin was saturated with Ba during this procedure. From the known Ba content of the concrete and the amount added it was calculated that for this method test with a complex sample matrix the loading capacity of the MnO_2 resin for Ba was 212 μg Ba / g resin.

The Ba content of the concrete in this case was 67 ppm (1 g of concrete was analysed). This has obvious repercussions for the use of MnO_2 resin for Ba enriched concretes, as the resin bed volume will have to be increased according to Ba content.

It was found that some Mn bleeds off the MnO_2™ resin column, hence the requirement for the Fe/Mn co-precipitation. This could be prevented by placing a Chelex™ column immediately below the MnO_2 column, to remove any Mn from the column eluent. This would add to the cost of the analysis however, and the since the Fe precipitation needs to be carried out anyway the Chelex™ column may be redundant.

The analytical results of this method were checked against a previous analysis of the ^{41}Ca bearing concrete, carried out using the method described by Suarez et al (2000) with the ^{41}Ca determination by both LSC and AMS. These comparisons yielded a result of 26.21 ± 2.97 Bq/g. The LOD of the method calculated using the method set out by Currie (1968) was 0.3 Bq/g (^{41}Ca), 0.03 Bq/g (^{45}Ca) and 0.04 Bq/g (^{90}Sr). These values are based on a sample mass of ~ 1g and a count time of 120 minutes.

The most obvious application of this method is in waste characterisation associated with nuclear decommissioning, where activated concrete bioshields are a potential source of $^{41/45}Ca$ activity. A core from a bioshield has been analysed using gamma spectrometry, $^{3}H/^{14}C$ analysis and $^{41/45}Ca$ determination. The core used was too old for any ^{45}Ca to be detected, but the ^{41}Ca activity showed an exponential decrease with distance away from the reactor. This was mimicked by other radionuclides such as ^{60}Co, ^{152}Eu and ^{3}H (figure 4). The exponential decrease in activity reflects the attenuation of the neutron flux, and consequently decreasing activation of the concrete components.

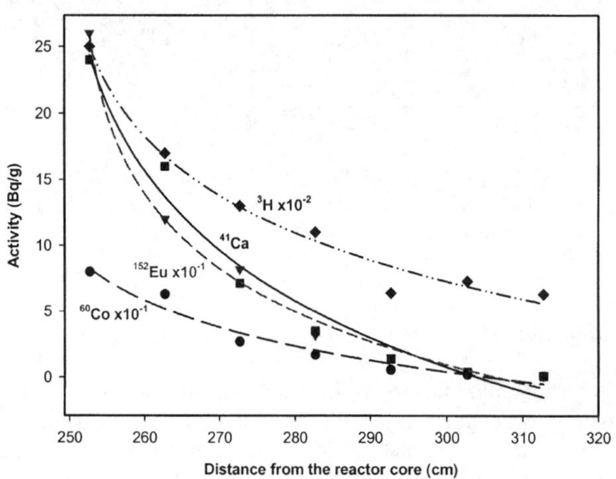

Figure 4. *Distribution of activity in activated concrete bioshield core.*

3 CONCLUSIONS

The method described provides a relatively rapid, cost effective method of determining $^{41/45}$Ca and ^{90}Sr in concrete cores. It performs at least as well as other published methods, with the additional benefit that all pH and concentration-dependant precipitations have been removed, making the analysis more robust and easier to carry out. The mass of MnO_2 resin can be adjusted depending on the Ba content of the concrete being analysed (determined by XRF along the with stable Ca determination).

References

A. Bombard (2005) MnO2 resin: A new approach for radium separation. Talk presented at Eichrom Users Meeting, Manchester.

L. A. Currie (1968). Limits of qualitative detection and quantitative determination. Analytical Chemistry, **40 (3),** 586-593

X Hou (2005) Radiochemical determination of 41Ca in nuclear reactor concrete. Radiochim. Acta **93**: 1-7

M. Itoh, K. Watanabe, M. Hatakeyama , M. Tachibana (2002) Determination of 41Ca in biological-shield concrete by low-energy X-ray spectrometry. Anal Bioanal Chem., **372**:532-536

J. Korkisch (1989) Handbook of ion exchange resins, Volume V. CRC Press

J. A. Suarez, A. G. Rodriguez , A. G. Espartero, G. Pina G (2000) Radiochemical analysis of 41Ca and 45Ca. Applied radiation and isotopes **52**: 407-413

THE CHEMISTRY OF ULTRA-RADIOPURE MATERIALS

HS Miley, CE Aalseth, AR Day, OT Farmer, JE Fast, EW Hoppe, TW Hossbach, KE Litke, JI McIntyre, EA Miller, A Seifert, GA Warren

Pacific Northwest National Laboratory, P.O. Box 999, Richland, Washington, USA 99352

1 INTRODUCTION

Ultra-pure materials are needed for the construction of the next generation of ultra-low level radiation detectors. These detectors are used for environmental research as well as rare nuclear decay experiments, e.g. probing the effective mass and character of the neutrino. Unfortunately, radioactive isotopes are found in most construction materials, either primordial isotopes, activation/spallation products from cosmic-ray exposure, or surface deposition of dust or radon progeny.

Copper is an ideal candidate material for these applications. High-purity copper is commercially available and, when even greater radiopurity is needed, additional electrochemical purification can be combined with the final construction step, resulting in "electroformed" copper of extreme purity. Copper also offers desirable thermal, mechanical, and electrical properties.

To bridge the gap between commercially-available high purity copper and the most stringent requirements of next-generation low-background experiments, a method of additional chemical purification is being developed based on well-known copper electrochemistry. This method is complemented with the co-development of surface cleaning techniques and more sensitive assay for both surface and bulk contamination. Developments in the electroplating of copper, assay of U and Th in the bulk copper, and the removal and prevention of residual surface contamination will be discussed relative to goals of less than 1 microBq/kg Th.

2 MOTIVATION

A germanium crystal must be kept cold and free from light to function as a gamma-ray spectrometer. This is usually accomplished with an Al or SS vacuum cryostat and numerous small parts for support, electrical connection, and conduction of heat to a liquid nitrogen reservoir. To lower the ambient background of such a spectrometer, enabling the detection and quantification of much lower quantities of radionuclides, these materials have been gradually replaced with copper and carefully selected plastics (see Figure 1). Further, the copper itself has undergone a transformation, from a common commercially supplied material to a carefully selected and chemically transformed material. The specific design of such a Ge spectrometer incorporates one or more kg of Cu structural material per kg of detector mass. Even low levels of contamination in this mass of copper could easily dominate the background for a specialized measurement such as double-beta decay of ^{76}Ge, in which the crystal provides both the sample and the measurement device.

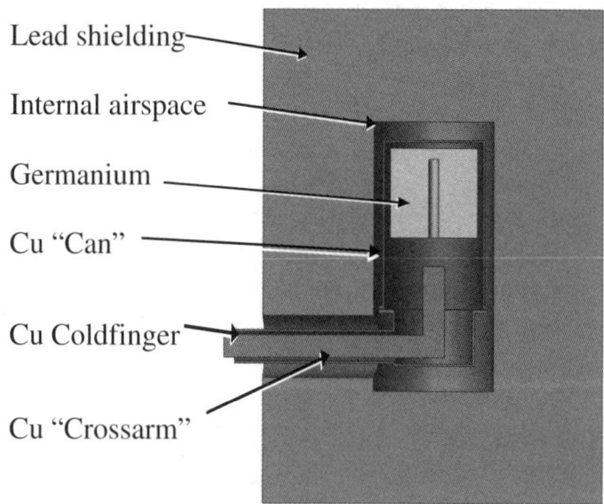

Figure 1 Design of a typical Ge spectrometer with about 2.7 kg of Cu per 1 kg of Ge

Frequently, cosmic-ray related signals in a low background spectrometer necessitate placement in an underground location. In addition, this system may require some consideration to the reduction of sample-associated backgrounds, either through decay, chemical separation, or coincidence techniques, listed in increasing order of expense.

The sensitivity of a Ge spectrometer will then depend on several factors, including depth and the purity of construction and shielding materials. Example background spectra obtained with several detectors are shown in Figure 2. However, it has recently been reported by Laubenstein that several systems' background spectra are not dependant on depth below perhaps 1000 meters water equivalent (mwe) of cosmic-ray shielding. Since the effect of cosmic-ray secondary muons is well known to continue decreasing with depth, this suggests that materials effects become dominant at that depth. In fact, it suggests that material backgrounds could be reduced by a factor of 1000 before the cosmic ray associated backgrounds would be dominant again at depths of 4000-5000 meters water equivalent.

As an example, an assay conducted several years ago[3] on electroformed copper presented ~8 kg of Cu to a specially produced germanium gamma-ray spectrometer at ~4000 mwe over a ~100 day period. Through observation of the ^{228}Th and ^{226}Ra daughters the authors concluded that if the ^{232}Th chain were in equilibrium, it would correspond to ~9 microBq/kg, and similarly, the ^{238}U chain concentration would have been <26 microBq/kg. A subsequent analysis by the authors has indicated[3,4] that the ~9 microBq/kg activity should be considered <9 microBq/kg based on the fact that the introduction of the 8 kg of Cu did not cause increased activity in the ^{228}Th chain, pointing to the activity being in small parts within the cryostat or surface activity on the Ge crystal itself. Thus, the key limits obtained by gamma-ray assay are <9 and <26 microBq/kg for the Th and U chains, respectively.

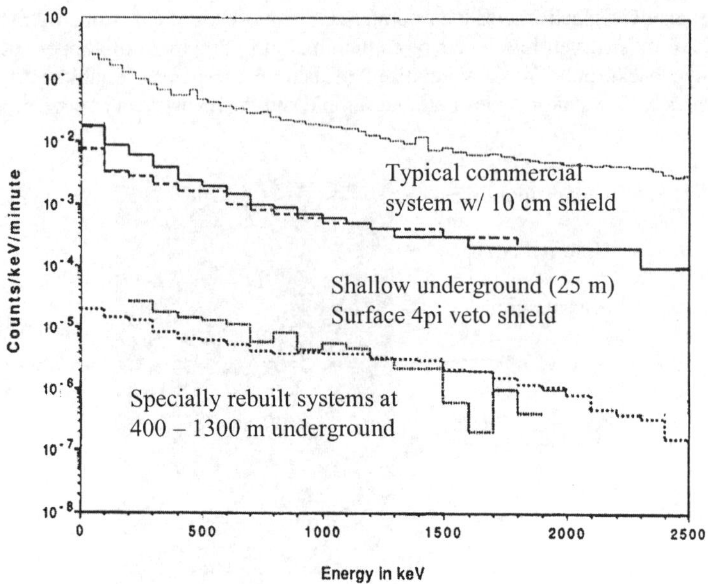

Figure 2 Comparison of spectra obtained by various systems.[2]

Other authors have recently reported different limits on different sources of Cu via gamma-ray spectroscopy, as shown in Table 1.

Table 1 Recently published gamma-ray-based assay limits for Th and U in electrolytic copper.

^{228}Th (^{232}Th)	^{226}Ra (^{238}U)	Comment
$< 1\times10^{-6}$ Bq/kg	$< 1\times10^{-6}$ Bq/kg	Goal of this research
$< 9\times10^{-6}$ Bq/kg	$< 26\times10^{-6}$ Bq/kg	In-house electroformed Cu discussed above
$< 28\times10^{-6}$ Bq/kg	$< 25\times10^{-6}$ Bq/kg	Commercially obtained electrolytic Cu Motta, et al.[5]
$< 12\times10^{-6}$ Bq/kg		Commercially obtained electrolytic Cu Rugel, et al.[6]
$< 19\times10^{-6}$ Bq/kg	$< 16\times10^{-6}$ Bq/kg	Commercially obtained electrolytic Cu Heusser, et al.[7]

While it may be tempting to draw conclusions about the quality of the various samples from the values in Table 1, it must be remembered that any of these materials may be orders of magnitude purer than these limits might suggest. However, these values do provide the introduction to an effort to document and achieve lower levels of contamination based on the ability to assay well below the limits afforded by gamma-ray spectroscopy.

3 ELECTROFORMED COPPER

Copper benefits from the "electrowinning" process used during refinement. This processing step electroplates the material from a sulfate solution onto cathodes. Most contaminants do not follow copper in this step by virtue of their negative electrochemical values as compared to copper, which has a cell potential of 0.342 V, and the resulting material, known as "electrolytic tough pitch" copper, is rather pure. Commercial high-purity copper usually undergoes further processing, including smelting to improve purity and hot rolling to improve density and mechanical properties.

Additional electrolytic and chemical purification can be combined with a final fabrication step, resulting in "electroformed" copper parts of extreme purity. In this process copper is electrodeposited onto forms, usually made of stainless steel in the shape of the desired final part. An example of an electroforming system is shown in Figure 3. This electroforming process can even be done underground, providing a potential way to eliminate cosmogenic activation products seen in copper that has had above-ground exposure, e.g. ^{60}Co.

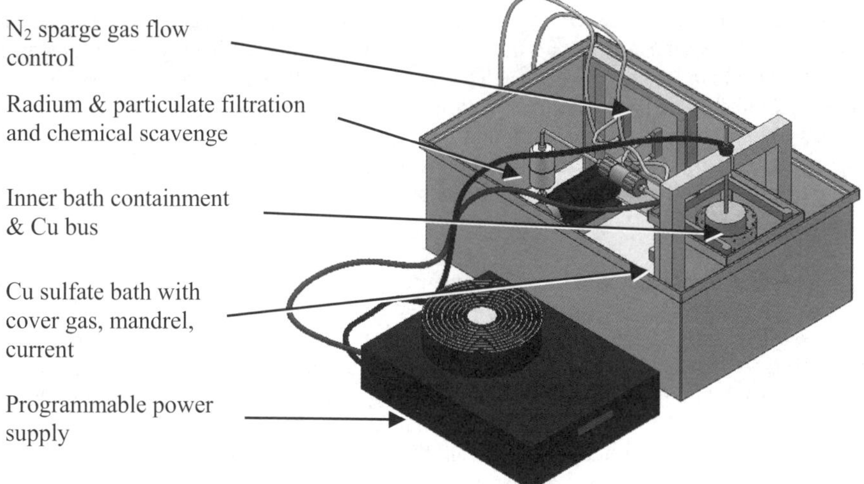

Figure 3 Schematic of the electroforming system.

3.1 Electrochemical Purity Considerations

The Nernst equation describes the tendency of an electrochemical reaction to go to the right, and can be used to model the behavior of ions in an electrochemical cell when that cell is in equilibrium.

For the generalized reaction

$$aA + bB + \ldots + ne^- \leftrightarrow cC + dD + \ldots, \tag{1}$$

the classic Nernst equation is given by

$$E_{cell} = E^0 - (RT/nF)*\ln([C]^c[D]^d/[A]^a[B]^b). \tag{2}$$

The more positive electrochemical values favor the formation of reduced species. For example, the Nernst equation predicts that thorium with a half cell potential of E^o Th = -1.9 V, would have to be at a concentration of over 10^{150} M before it would plate out at the half-cell potential of copper, E^o Cu = 0.34 V. Even when a reverse pulse plating process is employed to create level, and polycrystalline, copper, the Nernst equation predicts thorium would still have to be at a concentration of over 10^{100} M before it would plate out at -0.34 V. One would expect to obtain extreme copper purity from contaminants such as Th when electroplating at the voltages required for copper plating.

However, at high ionic strengths such as those used in most electroplating baths, activity coefficients cannot be calculated for the reactant and product species, leading to error in the Nernst-calculated value of E_{cell}. Also, as previously stated, the cell must be in equilibrium for the Nernst equation to hold. Any electrochemical cell under applied voltage is a dynamic system, where the concentrations of species are dramatically different surrounding the anode than they are surrounding the cathode. Ions are continuously being formed at the anode and reduced at the cathode. Given efficient mass transport, the concentration gradient is lessened, but the system remains in disequilibrium. Also, plating at rates of industrial practicality results in notable disequilibrium. A more accurate model would assume at least three different "zones" of equilibrium: one about the anode, one about the cathode, and one in the bulk of the plating bath. Ultimately, the presence of impurities in electroplated copper such as thorium which possess a very negative cell potential, indicates that the Nernst equation does not accurately model most real-world plating systems. In fact, it has been found by Hoppe, et al.[8] that at low concentrations in a copper sulfate bath, Th is only rejected at rates of 10^3 to 10^4, rather than 10^{100}. Obviously, mass transport and other factors dominate the behavior of contaminants regardless of redox potentials.

3.2 Cleaning Copper and Surface Treatments

Many low-background experiments which employ high-purity copper, e.g. Cuoricino,[9] have observed that surface contamination emerges as the dominant background. Radon daughters plate out on exposed surfaces, leaving a residual ^{210}Pb background that is difficult to avoid. Dust is also a problem; even under cleanroom conditions, the amount of U and Th deposited on surfaces can represent the largest remaining background.

Radiopurity is not the only motivation for surface cleaning; final machining processes also leave unwanted residues and copper fragments that make assembly difficult. Additionally, copper oxides can remain, creating unwanted additional infrared absorptive surface area. Surface cleaning is thus critical to ensuring high purity copper parts.

To address all these factors, an improved copper cleaning chemistry has been developed by Hoppe, et al.[10] Designed to replace an effective, but overly aggressive concentrated nitric acid etch, this peroxide-based method allows for more controlled cleaning of surfaces.

In an effort to keep the copper surfaces clean, passivation strategies have been developed to inhibit the reformation of oxides on the surface in an effort to minimize recontamination. The performance of the cleaning process and subsequent passivation were investigated by Hoppe, et al.[11]

4 PHYSICAL PROPERTIES OF ELECTROFORMED COPPER

4.1 Emissivity

Emissivity of both passivated and bare electroformed copper has been determined based on radiative thermal transfer. Electroformed copper cans at room temperature were fitted onto a cryostat, with a large thermal absorber (having a high emissivity coating), held near 80°K. Total heat transport was determined based on the rate of mass loss for the liquid nitrogen filled Dewar. Heat transport from the surface of interest was estimated by subtracting the heat transport of the cryostat alone (1.54 ± 0.06 W) from the total heating (ranging between 1.72 ± 0.05 W and 2.06 ± 0.07 W, depending on the surface of interest). Surface emissivity values were then extracted, using an analytic model for heat transfer between concentric convex shapes. Two samples of the passivated copper were examined, and were found to have emissivities, expressed as a percentage, of 2.7 ± 1.3 % and 2.8 ± 1.2 %, the stated errors representing statistical variations. These values are in reasonably good agreement with published values, which range from 1% up through 5% at room temperature, depending on preparation.[12-14] For bare cleaned copper, emissivity was estimated to be 8.0 ± 1.5 %. This is somewhat higher than published values; however, some small amount of oxidation or staining was observed, indicating that protocols to prevent exposure to oxygen may have been insufficient. The emissivity of a machined aluminum tube was also measured, and found to have a value of 5.3 ± 1.5 %, again in agreement with established values ranging from 3 to 10%.[14]

4.2 Mechanical Strength

The mechanical properties of electroformed copper can vary drastically depending on the conditions under which it was formed. Conditions that favor high purity can form large crystalline structures with poor mechanical strength. Small polycrystalline formations can exhibit adequate tensile strength but lower purities. Reverse pulse plating has allowed the growth of larger polycrystalline forms with the average hardness of 105-108 Vickers using 100 grams of force. This translates to approximately 300-340 MPa tensile strength.

5 STRATEGIES TO IMPROVE COPPER PURITY

Our current practice in preparing an electrochemical bath utilizes the highest purity acids commercially available to us. We have verified the concentration of contaminants stated by the vendor-supplied assay. We are also satisfied with the purity of water obtained from the output of laboratory grade ion exchange water treatment systems. We currently recrystallize all copper sulfate from saturated aqueous systems. The electrochemical baths are constantly circulated and filtered at 0.2 micron using a fluoropolymer cartridge. The bath is sparged with nitrogen boil-off from a stainless steel Dewar. This same nitrogen serves as a cover gas for the plating bath. A small amount of barium sulfate is loaded onto the filter to act as a radium scavenge.

In the future, we intend to perform the electroforming underground to limit the formation of cosmogenic species. We may need to take additional steps to improve the purity and quality of the copper such as electroforming our own anode material, which may also be formed underground. The exact waveform we use in the reverse pulse may be altered in order to offer an additional opportunity for those electronegative species, such as Th or U which may have been encapsulated during the forward portion of the plating process, to detach from the plate and be re-entrained into the bulk solution. Scrubbing of the bulk solution may be necessary using ion exchange, secondary electrogravimetric methods, or other processes. Use of the rejection rate information obtained from assay development work[8] will serve as a useful forecasting tool to determine what purity of copper can be expected from an electrochemical bath increasingly contaminated by the dissolution of anode material.

In order to determine if some of these improvements to the electroforming process are necessary, either additional rejection rate information must be obtained so that predictive models can be developed or more sensitive assays must be developed, which will be capable of measuring contaminants such as Th at our goal of less than 1 microBq/kg Th (0.25×10^{-12}g/g Cu). We plan to resume our work determining the rejection values using ^{228}Th as a tracer in the electroforming process.

6 ISSUES IN COPPER ASSAY

Concentrations of contaminants to meet our purity goal are far below the detection limits of many current analytical methods. As stated previously, radiometric assays require impractical amounts of material and exceedingly long counting times. Other analytical tools, such as Secondary Ion Mass Spectrometry (SIMS), lack adequate sensitivity or, such as Inductively-Coupled Plasma Mass Spectrometry (ICP/MS), lack the dynamic range to perform direct assays. ICP/MS, however, can detect many contaminants at or below these levels, including Th or U, if the bulk species of Cu can be reduced, since sensitivity of this assay is diminished by high copper concentration in the sample. Removal of the bulk species has been successful using ion exchange.[15] Unfortunately, the ion exchange resins are contaminated with the target species at levels that create significant background. We have confirmed the same contamination issues in our attempts to analyze samples using ion exchange to concentrate Th or U species from the bulk copper.

We have also pursued electrochemically back-plating of the copper sample to reduce the copper ion concentration and leave in solution impurities such as thorium and uranium, which should not plate out at the half-cell potential of copper.[8] Theoretically, the amount of sample that can be processed in this manner is not limited. All materials including any non-sample electrodes must not add contamination and must be of extreme purity. Also, the amount of copper remaining in solution must be back-plated to <10 μg/ml, and if a sulfate system is used, which is useful in support of further developing the predictive rejection rate information, then the sulfate ion should be <10 mmol as well. This approach hinges on the rejection rate remaining sufficiently high as to not introduce an undue amount of error. We have measured rejection rates as low as $\sim 10^3$ but even at 10^2 this would only represent a 1% error in the assay result.

7 CONCLUSIONS

In order to build a radiation detector capable of extreme sensitivity, historical copper electroforming has produced copper of < 9 microBq/kg ^{232}Th , and < 26 microBq/kg ^{238}U. However, future systems require levels below 1 microBq/kg. To achieve this, a new copper plating procedure has been devised and will be modified as needed to achieve the goal using ideas presented above. However, a copper assay capability below this level is desired. Limiting features of ion exchange plus ICP/MS approaches have been discussed. A new approach is being developed to produce samples of the copper to be assayed which are both low in Cu, to work within the dynamic range of the ICP/MS, and which have Th and U only from the original source of copper, not from the preparation materials or process. Fortunately, the current assay capability, 2-4 microBq/kg ^{232}Th, is not that far from the goal. Given that several useful reduction strategies exist, the authors feel it is likely that the sensitivity goal can be reached and that the rejection of Th and U by the electroforming process can be documented on a part-by-part basis. In any case, a one-time test to demonstrate Th rejection should be possible using ^{228}Th in the near future.

References

1 M. Laubenstein, et al., *Appl. Radiat. Isot.,*2004, **61**, 167.
2 H.S. Miley, et al., *J. Radioanal. Nucl. Chem., Articles,* 1992, **160** (2), 371.
3 R.L. Brodzinski, et al., *J. Radioanal. Nucl. Chem., Articles,* 1995, **193** (1), 61.
4 R.L. Brodzinski, private communications, 2006.
5 D. Motta, et al., *Nucl. Phys. B (Proc. Suppl.),* 2003, **118**, 451.
6 G. Rugel, et al., *Nucl. Phys. B (Proc. Suppl.),* 2005, **143**, 564.
7 G. Heusser, et al., *Proc. Intern. Conf. Isotop. Environm. Studies Aquatic Forum 2004*, 25 - 29 October 2004, Monte-Carlo, Monaco (http://edoc.mpg.de/220964).
8 E.W. Hoppe, et al., "Use of Electrodeposition for Sample Preparation and Rejection Rate Prediction for Assay of Electroformed Ultra High Purity Copper for ^{232}Th and ^{238}U Prior to Inductively Coupled Plasma Mass Spectrometry," submitted for publication to *J. Radioanal. Nucl. Chem.,* 2006.
9 C. Brofferioa, et al., *Nucl. Phys. B (Proc. Suppl.),* 2005, **145**, 268.
10 E.W. Hoppe, et al., "A method for removing surface contamination on ultra-pure copper spectrometer components," submitted for publication to *J. Radioanal. Nucl. Chem.,* 2006.
11 E.W. Hoppe, et al., "Cleaning and passivation of copper surfaces to remove surface radioactivity and prevent oxide formation," submitted for publication to *Nucl. Instrum. Methods Phys. Res. A,* 2006.
12 B. Window and G. Harding, *J. Opt. Soc. Am.,* 1981, **71** (3), 354.
13 D. Giulietti and M. Lucchesi, *J. Phys. D: Appl. Phys*., 1981, **14**, 877.
14 Infrared Services, Inc., http://www.infrared-thermography.com/material.htm.
15 P. Grinberg, et al., *Anal. Chem.,* 2005, **77** (8), 2432.

INDEPENDENT RADIOLOGICAL MONITORING; RESULTS OF A RECENT INTERCOMPARISON EXERCISE

K.S. Leonard[1], S. Shaw[2], N. Wood[3], J.E. Rowe[4], S.M. Runacres[3] D. McCubbin[1] & S.M. Cogan[1]

[1]Cefas, Lowestoft Laboratory, Pakefield Road, Lowestoft, Suffolk NR33 0HT, UK
[2]Harwell Scientifics, 551 South, Becquerel Ave, Harwell Int. Business Centre, Didcot, Oxon OX11 0TB, UK
[3]Food Standards Agency, Radiological Monitoring Branch, Emergency Planning, Radiation and Incidents Division, Aviation House, 125 Kingsway, London, WC2B 6NH
[4]Environment Agency, Lutra House, Dodd Way, Walton Summit, Bamber Bridge, Preston, PR5 8BX, UK

1 INTRODUCTION

The discharges of radioactive wastes, from nuclear sites in the United Kingdom, are authorised by the Environment Agency in England and Wales, under the Radioactive Substances Act, 1993[1]. The Food Standards Agency and the Environment Agency operate radioactivity surveillance programmes throughout the United Kingdom that are independent of the nuclear industry, the main aims being:

- To ensure that any radioactivity in food and the environment due to authorised radioactive releases and discharges do not compromise public health or the environment, undertaken by assessing critical group doses and comparing them to legal limits.
- To provide an independent check on monitoring data supplied by site operators.
- To provide reassurance that the radiological impact of authorised discharges of radioactive waste and other transfers of radioactivity into the environment is acceptable – incorporating monitoring related to less significant exposure pathways.
- To establish long-term information on concentrations and trends so that any changes can be quickly identified and action taken if required.

The combined results from the surveillance programmes are reported in the joint UK regulators' annual Radioactivity in Food and the Environment (RIFE) series of reports. The most recent report is RIFE 11, which provides all information for the monitoring carried out in 2005[2].

As part of an overall strategy for maintaining a quality data supply, the Environment Agency and the Food Standards Agency devised a small schedule of inter-comparison samples, from within their routine radiological surveillance programmes, to assess the output of radioanalytical results from each. Two of their contractors, Cefas and Harwell Scientifics, carried out sample collection, and subsequent radioanalysis, on behalf of the Food Standards Agency and the Environment Agency, respectively.

2 METHODOLOGY

The inter-comparison exercise commenced in 2005, in conjunction with the 2005 radiological surveillance programmes. The schedule of inter-comparison samples consisted of a number of sediment and seaweed samples, collected from selected sites (Sellafield and surrounding area, Trawsfynydd, Cardiff, Dungeness and Heysham) that are part of the on-going scheduled radiological surveillance programmes. A full description of the sampling details is given in Table 1, including site/location, sample type and sample frequency.

Table 1. *Sampling details of inter-comparison samples*

Sample Collector	Site/Location	Grid Reference	Sample Type	Sample Frequency
Cefas/HS[2]	Cardiff (Orchard Ledges East)	ST 213 747	Mud	Biannual
Cefas/HS[2]	Dungeness (Rye Harbour)	TR 943 190	Sandy mud	Biannual
Cefas/HS[2]	Heysham (Half Moon Bay)	SD 404 607	*Fucus V.*	Alternate quarters
HS[1]	Sellafield (Carleton Marsh)	SD 064 983	Sediment	Quarterly
HS[1]	Sellafield (Ravenglass Raven Villa)	SD 085 967	Sediment	Quarterly
Cefas[1]	Sellafield (Pipelines)	NY 018 033	Sand	Quarterly
HS[1]	Sellafield (St Bees West)	NX 959 116	Sediment	Quarterly
Cefas[1]	Trawsfynydd (East of pipe)	SH 696 383	Mud	Biannual

[1]Sampling by single contractor for both laboratories
[2]Independent sampling carried out by each laboratory

As indicated in Table 1, at five of the locations, either Cefas or Harwell Scientifics (HS) collected the required samples and provided aliquots for both laboratories to analyse. At the remaining three locations (Cardiff, Dungeness and Heysham), both Cefas and Harwell Scientifics collected samples independently for analysis (i.e. although within the same sample frequency period, samples were not collected at the identical time nor exactly the same location). This independent sampling was undertaken to represent and assess the likely variability that is inherent from the operation of independent surveillance programmes and work carried out by different contractors.

Having collected or received samples from each location, each laboratory prepared and analysed the samples in accordance with their own routine analytical methods using their own quality assured procedures i.e. no attempt was made to enforce the use of completely consistent methods or the use of single procedures by both laboratories. All of the samples were analysed by gamma spectrometry. In addition, samples collected at Heysham and Cardiff were analysed for ^{99}Tc and Total ^{3}H, respectively. Both laboratories use methods that are acceditated by the United Kingdom Accreditation Service (UKAS) to the ISO17025:2005 standard.

At Cefas, samples were assayed for gamma emitting radionuclides using high resolution p-type HPGe detectors, calibrated using a mixed radionuclide standard covering an energy range of approximately 60-2000 keV. The resultant spectra were resolved using proven standard gamma spectrometry software. Total ^{3}H analysis was achieved by wet oxidation using chromic acid to convert tritiated species to tritiated water. The tritiated water was purified (by distillation) and assayed by liquid scintillation counting. Samples were assayed for ^{99}Tc using a low background gas-flow proportional beta counter, following ashing under alkaline conditions, digestion with a mixture of hydrochloric acid and hydrogen peroxide, and chemical separation. Stable Re was used as a yield tracer.

At Harwell Scientifics, the procedures used to assay gamma emitting radionuclides were similar to those described for Cefas. In contrast, the analysis of both tritium and ^{99}Tc by Harwell Scientifics were undertaken by slightly different approaches. For Total ^{3}H analysis, a combustion technique (instead of wet oxidation) was used to convert tritiated species to tritiated water. The combustion technique involves gradually heating samples in an oxygen-rich atmosphere in the presence of a copper oxide catalyst. The tritiated water was selectively trapped in a series of gas bubblers containing dilute acid and assayed by liquid scintillation counting. For ^{99}Tc analysis, the samples were ashed under alkaline conditions and digested in nitric acid. Purification was carried out chemically followed by ion chromatography separation. ^{99}Tc assay was achieved using liquid scintillation counting (as an alternative to beta counting) and using ^{99m}Tc as a yield tracer.

3 RESULTS AND DISCUSSION

3.1 Comparison of gamma emitting nuclides (^{137}Cs, ^{60}Co and ^{241}Am)

From a number of possible gamma emitting nuclides, 3 specific nuclides were chosen to compare the observed radioanalytical data, ^{60}Co (high energy, ~ 1332kev), ^{137}Cs (mid energy, ~ 662kev) and ^{241}Am (low energy, ~ 60kev). Results for these gamma emitting radionuclides are provided in Figure 1 for ^{60}Co, ^{137}Cs and ^{241}Am.

The magnitude of activity concentrations, observed across the range of inter-comparison samples, varied widely. Therefore, the resultant data have been plotted on a logarithmic scale to aid presentation. For sampling undertaken by one organisation (and the sample split between the two laboratories), the concentrations spanned two orders of magnitude over the whole of these sites. Overall good consistency is observed for each of the three reported radionuclides. One set of data was marginally lower than the other, with the differences between the values for ^{60}Co and ^{137}Cs being minimal (average of just 4% and 1%, respectively), and are well within uncertainty budgets. The discrepancy between the overall results for ^{241}Am was greater (average of 22%); the increased variation may be due to slight differences in the methods used to correct for self absorption (at low energy). It is also worth noting that differences of similar magnitude are expected for the comparison of gamma spectrometry measurements with ^{241}Am data obtained by alpha spectrometry methods.

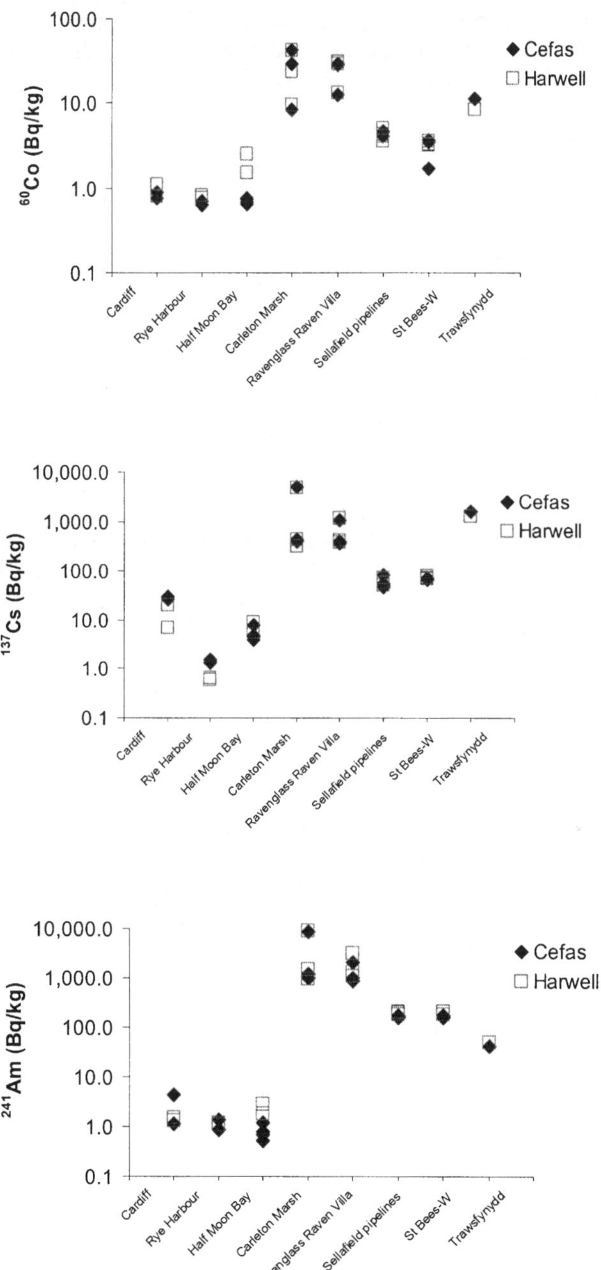

Figure 1. *Comparison of radionuclide concentrations from sampling and analysis by two independent contractors (^{60}Co, ^{137}Cs and ^{241}Am)*

It is particularly reassuring to note that both laboratories observed almost identical trends in data for the mud samples collected from Carlton Marsh (in the vicinity of Sellafield). The sample collected in the last quarter contained concentrations of ^{137}Cs and ^{241}Am that were elevated (by an order of magnitude) compared with those obtained earlier in the year. In contrast, concentrations of ^{60}Co were noticeably reduced (by ~ 4 fold). A similar, less marked, trend was apparent in the results for sandy mud obtained from the Ravenglass estuary (concentrations of ^{137}Cs and ^{241}Am in the last quarter elevated by ~3 fold whilst the ^{60}Co concentration was halved). At both sites, the temporal variation in activity concentrations was attributed to the sampling, as opposed to any effect from recent Sellafield discharges. It is likely, due to the dynamic environment in which the samples were collected, that the samples from the last quarter consisted primarily of sub-surface material, with contamination typical of greater historic ^{137}Cs and ^{241}Am discharges from Sellafield. In contrast, data for the other samples appear typical of surface sediment contaminated by more recent Sellafield discharges (hence higher concentrations of ^{60}Co but reduced concentrations of ^{137}Cs and ^{241}Am).

Data to assess the compatibility of results provided by independent sampling are more restricted, being limited to just two samples per annum from three sites (Cardiff, Dungeness and Heysham). In addition, interpretation of the available results is complicated by the fact that activity concentrations in these samples were very low in comparison to the other sampling sites. More specifically, concentrations of ^{60}Co and ^{241}Am were below detection, in all but the *Fucus Vesiculosus* (seaweed) samples obtained from Half Moon Bay in the vicinity of Heysham. One laboratory observed ^{60}Co and ^{241}Am concentrations in range of 0.6 - 0.7 Bq kg^{-1} and 0.7 - 1.1 Bq kg^{-1}, respectively, and are in agreement with reported values of <3 Bq kg^{-1} (for both nuclides) by the other laboratory. ^{137}Cs concentrations in seaweed from Half Moon Bay and in sandy mud from Rye Harbour (in the vicinity of Dungeness) were low. Reported values from each laboratory were in the range of 3.9 - 7.8 Bq kg^{-1} compared with 5.5 - 8.9 Bq kg^{-1} for Half Moon Bay, and in the range of 1.4 - 1.6 Bq kg^{-1} compared with <0.6 Bq kg^{-1} for Rye Harbour. ^{137}Cs concentrations in mud from Orchard Ledges (in the vicinity of Cardiff) were higher in comparison to the aforementioned locations. Reasonable agreement was observed between laboratories' data for one of the samples (20 Bq kg^{-1} compared with 26 Bq kg^{-1}) but less so for the other sample (7 Bq kg^{-1} compared with 30 Bq kg^{-1}). However, the samples were taken ~ 3 months apart and the effects of temporal variation are likely being observed here.

3.2 Comparison of Total ^{3}H data (Cardiff)

Data for total tritium activity concentrations in Orchard Ledges mud, obtained by independent sampling in the vicinity of the GE Healthcare plc radiopharmaceutical plant at Cardiff, is provided in Table 2.

With the exception of one analysis, which appears to be slightly low, the other Total ^{3}H values appear consistent with an average activity of 98 ± 19 Bq kg^{-1}. As noted previously for the ^{137}Cs data at this location, a possible explanation for the "low" value is that the comparison with its counterpart is influenced by the time interval between the two sample collections.

Table 2. *Comparison of total ^{3}H activity concentrations in Orchard Ledges Mud (Cardiff) arising from independent sampling and analysis*

Collection date	Sample	Contractor	Result (Bq kg^{-1} wet)[1]
24/01/2005	EA3365	HS	50 ± 11
21/04/2005	2005000295	Cefas	91 ± 19
16/08/2005	EA4045	HS	120 ± 20
17/08/2005	2005000847	Cefas	82 ± 17

[1] ± figures quoted are total method uncertainties

3.3 Comparison of ^{99}Tc data (Heysham)

Results for the analysis of ^{99}Tc activity concentrations in Half Moon Bay Seaweed, obtained by independent sampling in the vicinity of the nuclear power plant at Heysham are provided in Table 3.

Table 3. *Comparison of ^{99}Tc concentrations in Half Moon Bay Seaweed (Heysham) arising from independent sampling and analysis*

Collection date	Sample Identifier	Contractor	Result (Bq kg^{-1} wet)[1]
18/03/2005	EA3573	HS	580 ± 40
13/04/2005	2005000317	Cefas	1010 ± 106
11/07/2005	2005000652	Cefas	1080 ± 114
22/07/2005	EA3496	HS	550 ± 40

[1] ± figures quoted are total method uncertainties

Concentrations in both the samples collected by one laboratory (average of 1045 Bq kg^{-1}) were almost 2 fold greater than those obtained by the other (average of 565 Bq kg^{-1}). There is potential for the two laboratories to produce inconsistent ^{99}Tc seaweed data for a number of reasons including the sampling of young or old thallus, as well as the analytical methods being different. In addition, uptake in seaweed is known to be variable and dependant upon local conditions at the time of sampling. The principal source of inconsistency of radionuclide concentrations in seaweed is the variation in the concentration in the surrounding water[3]. Further work is required to confirm whether or not the apparent discrepancy is indeed 'real', or the result of scatter in the data produced by environmental variation. Although the most appropriate reason is not fully evident, the variation is not considered excessive for the purpose of the UK surveillance programmes.

4 CONCLUSIONS

The conclusions are summarised as follows:

- Results for concentrations of gamma emitting nuclides for samples that had been collected by a single contractor (on behalf of both laboratories) were reasonably consistent, with activity concentrations spanning two orders of magnitude over the whole of these sites.

- Both laboratories observed relatively enhanced concentrations of ^{137}Cs and ^{241}Am in two sediment samples (in the last quarter samples) obtained from two sites within the vicinity of Sellafield – It is likely that the samples collected in the last quarter consisted primarily of sub-surface material, with contamination typical of greater historic ^{137}Cs and ^{241}Am discharges from Sellafield.
- Results for concentrations of gamma emitting radionuclides, using independent sample collection, were reasonably consistent, however with the caveat that there was limited amount of data and activity concentrations were either close to or below the level of detection.
- Results for Total ^{3}H, using independent sample collection, were in reasonable agreement given the likely environmental variability.
- Some variation was observed for ^{99}Tc results. Uptake of ^{99}Tc by seaweed is known to be variable and dependant upon the local conditions at the time of sampling.
- The comparison work also helps to provide a baseline for the monitoring programmes in the event of changing analytical laboratories, or the methods being used for analysis. Results can be assessed against the spread of data from the two laboratories, for the sites compared (as well as against past data sets).
- Overall, the results provide the Food Standards Agency and the Environment Agency with confidence in the compatibility of the monitoring data produced by their laboratories.

Acknowledgements

The Environment Agency and Food Standards Agency funded this work as part of their respective radiological surveillance programmes. The authors would also like to express their sincere thanks to our colleagues at Cefas and Harwell Scientifics for their contributions to the collection, distribution, preparation and radioanalysis of samples.

References

1. United Kingdom - Parliament, 1993. Radioactive Substances Act, 1993. HMSO, London.

2. Radioactivity in Food and the Environment, 2005. RIFE 11, (2006). Environment Agency, Environment and Heritage Service, Food Standards Agency and Scottish Environment Protection Agency. Bristol, Belfast, London and Stirling.

3. C. Nawakowski, M.D. Nicholson, P.J. Kershaw and K.S. Leonard. J. Environ. Radioactivity, 2004, 77, 159-173.

ROUTINE APPLICATION OF CN2003 SOFTWARE TO LABORATORY LIQUID SCINTILLATION CALIBRATION

P.E. Warwick, I.W. Croudace & N.G. Holland

GAU-Radioanalytical, National Oceanography Centre, European Way, Southampton SO14 3ZH, UK

1 INTRODUCTION

The accelerated decommissioning programme in the UK has generated a demand for the characterisation of a wide range of radionuclides in diverse waste forms. A significant proportion of these radionuclides are pure beta emitters or electron capture radionuclides which are typically quantified using liquid scintillation counting. For some of these radionuclides, no traceable standards exist which are suitable for instrument calibration and proxy radionuclides with similar decay modes and energies are often used for calibration. Such an approach relies on the appropriate selection of a proxy radionuclide that is sufficiently similar in terms of beta decay energy, the type of transition and the beta shaping factor. Often just the beta energy is considered and this may lead to inaccurate efficiency predictions for the nuclide of interest. A potentially more robust approach is to theoretically determine the efficiency for a radionuclide based on fundamental parameters. Such an approach has been developed as the Ciemat Nist (or CN) method which is often used for standardisation of beta emitting radionuclides under carefully controlled conditions (Günther, 2002 and references therein). In this study, the application of the method to routine radioanalytical calibrations is evaluated.

The computer program CN2003 combines a number of individual programs developed for the prediction of liquid scintillation counting efficiencies for pure beta, beta/gamma and electron capture radionuclides. The program offers an attractive alternative to certified standards for liquid scintillation calibration as well as for wider applications including independent confirmation of instrument calibrations, calibration for short lived radionuclides and checking of standard solutions. In this study, the routine application of the CN2003 software for liquid scintillation calibration was evaluated, comparing the predicted efficiency curves for a range of radionuclides with experimentally determined calibrations using traceable standardised radionuclide solutions, with consideration of the impact of source composition and quenching agent. In addition, the validity of using proxy radionuclides for calibration where no traceable standards currently exist was assessed.

2 METHODOLOGY

A Wallac 1220 Quantulus liquid scintillation counter was used for all measurements. The instrument was configured for low energy bias with PSA/PAC options disabled. A full-energy counting window was used for all measurements. Certified radionuclide solutions were purchased from Amersham-QSA and were diluted gravimetrically prior to use. All sources were prepared in 22ml polythene scintillation vials (Meridian, Epsom, UK). Gold Star cocktail was supplied by Meridian, Epsom, UK and Ultima Gold AB was supplied by Perkin Elmer, UK. All other reagents were supplied by Fisher Scientific Ltd, Loughborough, UK.

2.1 Predicted vs measured quench curves

A range of radionuclides was chosen including pure beta emitters, beta/gamma emitters and electron capture radionuclides. These radionuclides were used to prepare single nuclide sets of calibration standards with a matrix typical of that routinely encountered (Table 1). The quench level was varied by altering the ratio of aqueous fraction to cocktail whilst maintaining the total volume of liquid in the vial. The set of calibration standards were dark-adapted and then counted to determine the measurement efficiency over a range of quench levels.

Table 1. *Details of radionuclides considered in this study*

Nuclide	Decay mode (E_γ or $E_{\beta max}$)	Aqueous phase	Scintillant	Total volume (ml)
^{14}C	β (157 keV)	Carbosorb	Gold Star	20
^{36}Cl	β (710 keV) EC	7M NH_4OH	Gold Star	20
^{41}Ca	EC	3M HCl	Gold Star	20
^{45}Ca	β (257 keV)	3M HCl	Gold Star	20
^{55}Fe	EC	2M H_3PO_4	Ultima Gold AB	17
^{63}Ni	β (65.9 keV)	1.2M HCl	Gold Star	20
$^{90}Sr/^{90}Y$	β (546 + 2279 keV)	1.2M HCl	Gold Star	20
^{137}Cs / ^{137m}Ba	β (512 keV) IT (661 keV)	1.2M HCl	Gold Star	20
^{147}Pm	β (224 keV)	1.2M HCl	Gold Star	20
^{241}Pu	β (20.8 keV)	1.2M HCl	Gold Star	20

The experimentally determined quench curve was compared with the theoretical quench curve predicted using the CN2003 software. For all predictions, a kB value of 0.0075 cm^2MeVg^{-1} was chosen. The program 'KB' was used for ionisation quenching approximation and 'XCOM' selected for cross section approximations. The CN2003 program predicts the measurement efficiency of the radionuclide relative to that of ^3H at a given quench level. The actual ^3H efficiency was measured for a range of quench levels in matrix matched sources and these results were used to determine the ^3H counting efficiency for any quench level by interpolation. The absolute efficiency for the radionuclide of interest was then calculated for a given quench level by combining the ^3H

quench curve with the predicted radionuclide efficiency relative to ^{3}H (Figures 1 – 3). For ^{90}Sr, it was assumed that ^{90}Y was in equilibrium with ^{90}Sr. The counting efficiencies for ^{90}Sr and ^{90}Y were determined separately and then summed to give the combined counting efficiency.

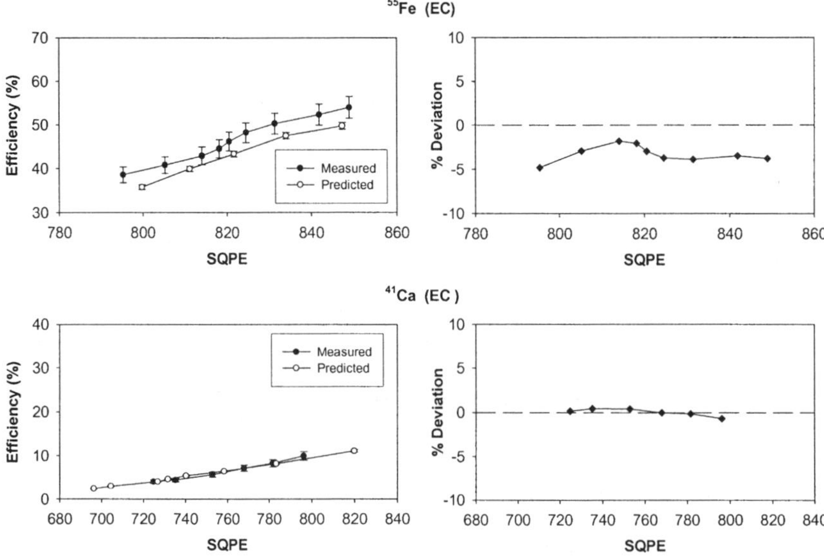

Figure 1 *Measured vs predicted efficiencies for radionuclides decaying by electron capture*

2.2 Use of proxy radionuclides

Where certified standards of a radionuclide are unavailable for LSC calibration, a proxy radionuclide with a similar beta energy end point is often used. In this study, ^{79}Se, ^{93}Zr and ^{151}Sm were chosen as radionuclides for evaluation as they are typically calibrated using this approach. For ^{79}Se (E_{max} = 151 keV), ^{14}C (E_{max} = 157 keV) was chosen as the proxy radionuclide, whilst ^{63}Ni (E_{max} = 66 keV) was chosen for both ^{93}Zr (E_{max} = 61 keV) and ^{151}Sm (E_{max} = 76 keV). The predicted efficiencies for the selected radionuclides were determined using the CN2003 software and compared with that for the appropriate proxy radionuclide (Figure 4).

3 RESULTS AND DISCUSSION

The general agreement between predicted and measured quench curves is very good with all predicted values being within 5% of the measured values. For the electron capture nuclide ^{41}Ca and pure beta emitting radionuclides ^{14}C, ^{45}Ca, ^{63}Ni, $^{90}Sr/^{90}Y$, ^{147}Pm and ^{241}Pu the discrepancy between theoretical and measured efficiencies was <1% demonstrating the effectiveness of the theoretical prediction over a wide range of energies. Similar good agreement was also observed for the beta/gamma emitting radionuclide, ^{137}Cs

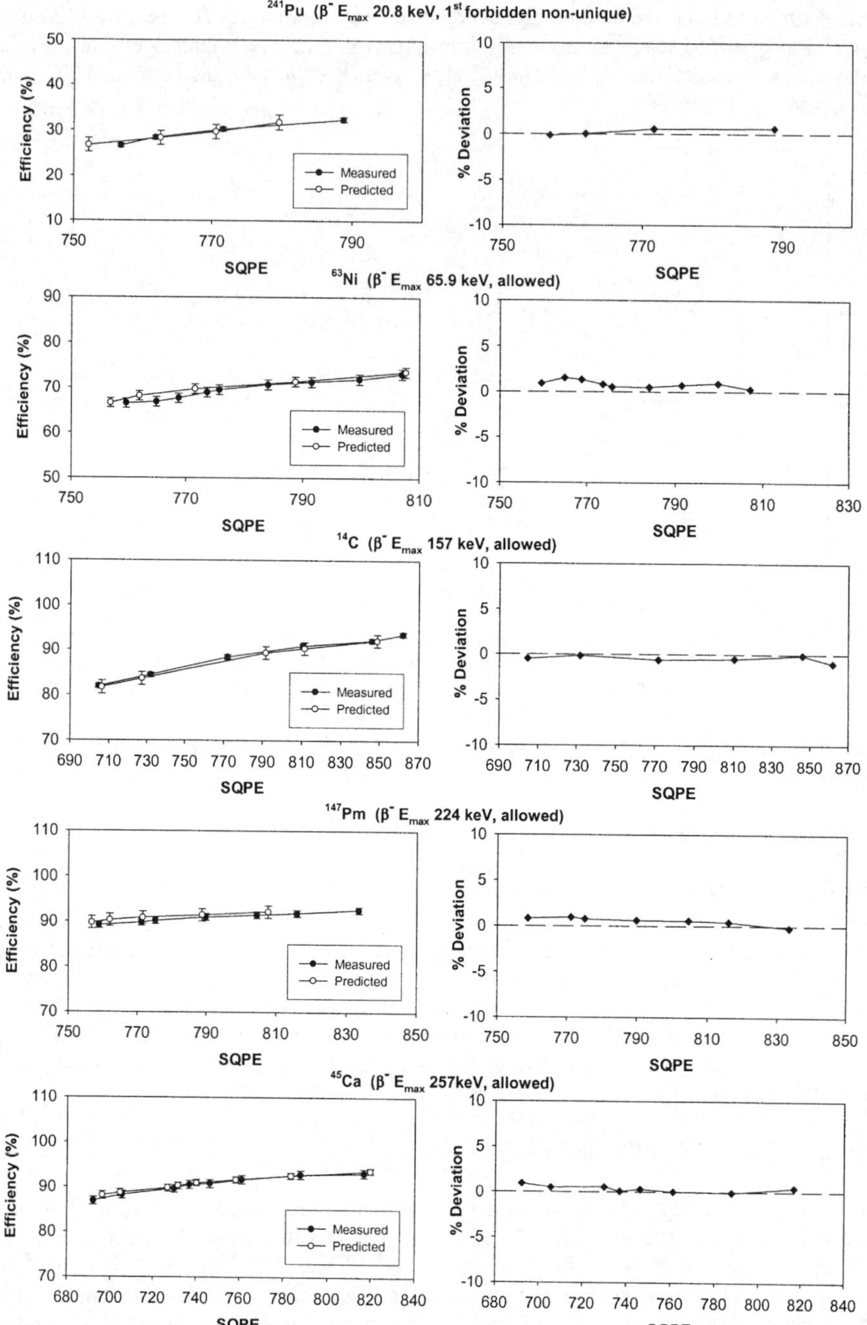

Figure 2 *Measured vs predicted efficiencies for radionuclides decaying by low energy beta emission*

demonstrating that the program was effectively accounting for gamma and isomeric transition interactions. The greatest discrepancies were observed for the electron capture radionuclide ^{55}Fe and the beta/EC nuclide ^{36}Cl. This may be related to limitations to the models used by CN2003 to predict efficiencies for electron capture radionuclides (Günther, 2002) and in particular the choice of stopping powers, kB, (Grau Carles et al, 2004) or the composition of the sources not being accurately modelled. For ^{36}Cl, the discrepancy may also be associated with the potentially inappropriate use of a constant beta shape factor (Grau Carles, 1995). Within the CN2003 model, the choice of cocktail type, sample composition and volume did not significantly alter the predicted efficiencies. The choice of kB value did impact on the predicted efficiencies particularly for ^{55}Fe where deviations of up to +1.5% were observed. In all instances, the predicted quench curve is sufficiently precise for routine radionuclide quantification.

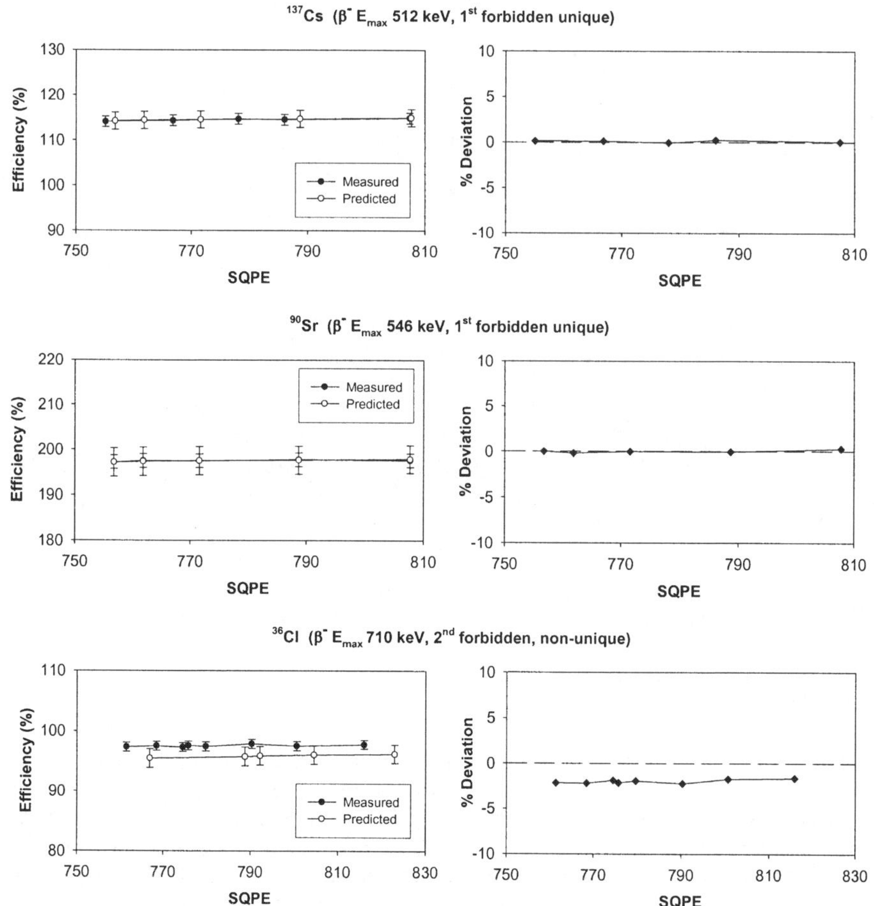

Figure 3 *Measured vs predicted efficiencies for radionuclides decaying by high energy beta emission and beta/gamma emission.*

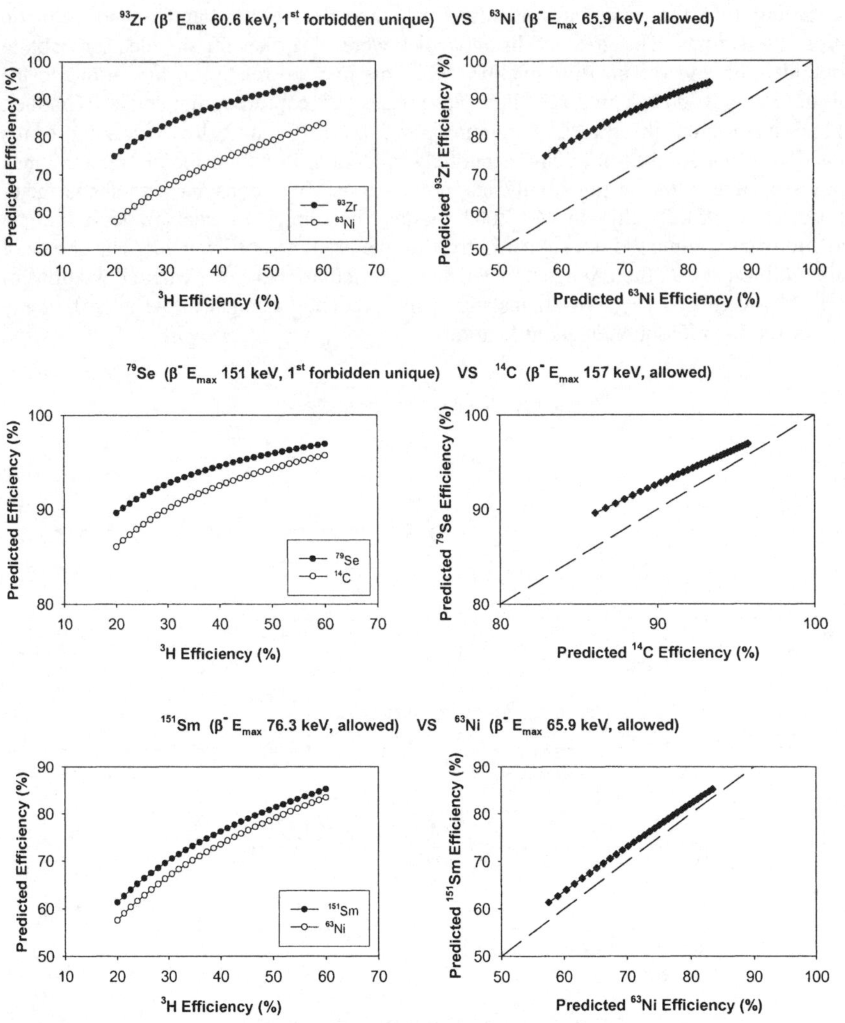

Figure 4 *Comparison of radionuclide efficiencies with those of common proxy radionuclides.*

Application of the CN2003 software to evaluate the suitability of proxy radionuclides for LSC calibration in the absence of certified radionuclide standards is also of considerable benefit. Comparison of predicted efficiency curves for ^{151}Sm/^{63}Ni confirms that the uncertainties introduced through the use of proxy radionuclides are insignificant. For ^{79}Se, the deviation in efficiency from that of the proxy radionuclide (^{63}Ni) increased with increasing quench although the maximum deviation was still <5%. However, for ^{93}Zr/^{63}Ni, the discrepancy of up to 15% is significant. Further research is required to identify the cause of this discrepancy and to assess whether the choice of proxy is appropriate in this case.

4 CONCLUSIONS

The CN2003 software program offers an effective approach to the calibration of liquid scintillation counters for routine radioanalytical applications. Agreement between measured and theoretical efficiencies was good over a wide quench range and for varying sample compositions. Discrepancies between measured and predicted efficiencies for beta, beta/gamma and electron capture radionuclides are typically $<5\%$ and in most cases is $<1\%$. The most notable discrepancies were observed for ^{55}Fe and ^{36}Cl although even in these cases, the discrepancy was *ca* 5% and still acceptable as long as the uncertainties are suitably propagated. The CN2003 program could therefore be applied to the routine calibration of liquid scintillation counters either where no certified standard exists (avoiding the need for proxy radionuclides) or where a radionuclide is sufficiently short lived that repeated purchase of certified standards is undesirable or uneconomic.

References

A. Grau Carles (1995). New methods for the determination of β-spectra shapefactor coefficents. *Appl. Radiat. Isot.*, ***46***, *125-128.*

A. Grau Carles, E. Günther, G. Garcia & A. Grau Malonda (2004). Ionisation quenching in LSC. *Appl. Radiat. Isot.*, ***60***, *447-451.*

E. Günther (2002). What can we expect from the CIEMAT/NIST method? *Appl. Radiat. Isot.*, ***56***, *357-360.*

EASY METHOD OF CONCENTRATION OF STRONTIUM ISOTOPES FROM RADIOACTIVE AQUEOUS WASTES FOR THE DETERMINATION OF ^{90}SR BY LIQUID SCINTILLATION COUNTING. APPLICATION OF STRONTIUM EMPORE™ RAD DISKS

E. Minne, F. Heynen, S. Hallez

Scientific Institute of Public Health, Section Radioactivity, rue J. Wytsman 14-16, 1050 Brussels, Belgium

1 INTRODUCTION

This study emphasizes the measurement procedure of ^{90}Sr activity in radioactive aqueous waste produced by the nuclear industry. Considering radioecological and radioprotective elements, there is a need for systematic measurements of ^{90}Sr activity in rejected wastes. The classical methods for this determination are gas flow proportional counting, liquid scintillation counting or Cherenkov counting.[1,2,3] These techniques require efficient concentration and purification steps before the measurement of radiostrontium itself.

In the lab, chromatographic extraction resin Eichrom™ Sr-spec in 2 mL columns are routinely used to determine the ^{90}Sr activity in natural water and foodstuffs. For a volume of water sample greater than 100 mL, it is essential to preconcentrate the alcalino-terrous elements. The different stages to obtain the purified solution of strontium require at least one or two days of handling. The aim of this study was to apply another technique in order to avoid the time-consuming strontium preconcentration step.

3M Corp. developed the Empore™ Strontium Rad Disks allowing the fixing of strontium by simple filtration of the water sample. No preconcentration step is required. Samples are acidified in 2-4 M nitric acid. Recoveries greater than 85 % have been observed for filtered volumes up to 3 L.[4,5] These disks contain a strontium-selective crown ether extractant bonded to a solid silica support inserted in an inert PTFE matrix.[6,7]

The recommended procedure RP515 from U.S. D.O.E. proposes to count the membrane by gas flow proportional counter or by liquid scintillation.[5] This approach was not followed. It was preferred to eluate the fixed strontium from the disk using a solution of disodic EDTA. The reason of this approach was a concern for an eventual interference of the solid matrix (filter) in the scintillating cocktail.

This paper describes the method and the necessary validation criteria to achieve the accreditation in accordance with Belgian standard NBN EN ISO/IEC 17025.[8] Several samples of radioactive aqueous effluents were analyzed in parallel by the method of the 3M Empore™ Sr Rad Disks described here and by the method using the Eichrom™ Sr-spec columns. The results of the two methods are compared at the end of this document.

2 ^{90}Sr DETERMINATION IN WASTES USING 3M EMPORE™ Sr RAD DISKS

2.1 Instrumentation

The manifold filtering device allows for simultaneous filtration of six samples simultaneously. Glass vials are placed under each filter support to collect the eluate. A vacuum pump applies a negative pressure to the water sample.

The counting is performed by an ultra-low level liquid scintillation counter 1220 Quantulus (Wallac™).

3M Empore Strontium Rad Disks (47 mm of diameter) were used to concentrate the strontium contained in the samples. The membrane eluates were mixed with scintillating cocktail Optiphase Hisafe 3 in 20 mL polyethylene vials from PerkinElmer™.

2.2 Separation of ^{90}Sr

The developed method generally follows the recommendations of the method RP515 from the U.S. D.O.E.[5] A known volume of sample (500 mL) is acidified to 2 M HNO_3. ^{85}Sr tracer in 0.1M HCl – 1mM $Sr(NO_3)_2$ is added to the sample to determine strontium recovery (activity between 10 and 20 Bq). Samples are filtered. Disk rinsing with 2 M HNO_3 and dionized water eliminates possible interfering radioisotopes. ^{90}Sr is then eluted with 6 mL of 0.025 M sodium EDTA (pH=11). 5 mL of the 6 mL of eluate are transferred in a liquid scintillation vial. These are counted at ^{90}Sr/^{90}Y equilibrium (about 3 weeks) after scintillation cocktail adding.

2.3 Sample preparation for counting

The applied counting sequence is as follows :
blank / standard ^{85}Sr / standard ^{90}Sr / samples (max. 6).
The composition of the different vials is described here :

- *blank* : 5 mL EDTA + v mL 0.1M HCl + 15 mL cocktail ;
- *standard ^{85}Sr* : 5 mL EDTA + v mL ^{85}Sr + 15 mL cocktail ;
- *standard ^{90}Sr* : 5 mL EDTA + v mL ^{90}Sr + 15 mL cocktail ;
- *samples* : 5 mL eluate + v mL HCl 0.1M + 15 mL cocktail.

"v mL" is the volume of ^{85}Sr tracer added to the sample (0.1 or 0.2 mL in function of the volumic activity of the tracer solution). This same volume "v mL" of 0.1 M HCl is added to the blank and the eluates so that conserving the same conditions of pH as the standards.

2.4 Counting informations

Blank and standards solutions are used to determine counting efficiencies in the different regions of interest (ROI) for each series of samples. The first ROI (W1) is located on the ^{85}Sr peak outside of the low energies area (see Figure 1). The second one is located on the peak of the ^{90}Y (W2), factor of merit being maximum around this peak. Indeed, the contribution of the ^{85}Sr is negligible in these high energies.

All samples are counted during ten hours. Moreover, it's chosen to wait secular equilibrium between ^{90}Sr and ^{90}Y before counting (approximately 3 weeks).

2.5 ^{90}Sr activity calculation

The chemical yield (η_{ch}) of the complete handling of the sample is defined in the equations (1) and (2) :

$$\eta_{ch} = \frac{R_{net-1}}{R_{th-1}} = \frac{R_{net-1}}{A_{85} * \eta_{85-1} * 60} = \frac{(R_1 - R_{blk-1}) - [C_{90-1}]}{A_{85} - \eta_{85-1} * 60} \quad (1)$$

$$\eta_{ch} = \frac{(R_1 - R_{blk-1}) - \left[\frac{(R_2 - R_{85-2})}{\eta_{90-2}} * \eta_{90-1} \right]}{A_{85} * \eta_{85-1} * 60} \quad (2)$$

The count rates (R) are given in CPM. This chemical yield is the ratio between net count rate due to the ^{85}Sr measured in the sample (R_{net-1}) and the theoretical count rate due to ^{85}Sr tracer initially added (R_{th-1}). The activity A_{85} is the activity of the tracer at measurement time. The counting efficiencies of ^{85}Sr in W1 (η_{85-1}) and in W2 (η_{85-2}) are calculated from the spectra of the standard ^{85}Sr and the blank (background).

The gross counting rate of the sample in W1 (R_1) contains counts due to ^{85}Sr and two contributions: firstly, background measured in the blank vial (R_{blk-1}) and secondly, counts due to ^{90}Sr and ^{90}Y (C_{90-1}), which have to be deduced. C_{90-1} will be established by means of the counting efficiencies of couple ^{90}Sr/^{90}Y in the two windows (η_{90-1} and η_{90-2}) and of the sample net counting rate in W2 ($R_2 - R_{85-2}$) (without the ^{85}Sr contribution).

The ^{90}Y activity (A_{90}) depends on the net counting rate of the sample in W2 ($R_2 - R_{85-2}$). It takes into account the counting efficiency of ^{90}Y in W2 (η_{90-2}) and the chemical yield of analysis (η_{ch}). When secular equilibrium is reached, the activity per unit of volume V (or a mass, if necessary) will be calculated as follows :

$$A_{90} = A_{Y-90} = A_{Sr-90} = \frac{R_2 - R_{85-2}}{\eta_{90-2} * \eta_{ch} * 60 * V} \quad (3)$$

Before the equilibrium :

$$A_{90} = A_{Sr-90} = \frac{A_{Y-90}}{\left(1 - e^{-\lambda_{Y-90} * t}\right)} \quad (4)$$

where A_{Y-90} is calculated as in (3), λ_{Y-90} is the radioactive decrease constant of yttrium-90 and t is the time passed between the end of sample elution and the moment of measurement.

3 ^{90}Sr DETERMINATION IN WASTE USING EICHROM^{TM} Sr-SPEC CHROMATOGRAPHIC RESIN

The determination of ^{90}Sr activity in natural waters and in foodstuffs by liquid scintillation is accredited by the Belgian Accreditation Office (BELAC). It follows Eichrom^{TM} procedure SRW 01 Rev. 1.4.[9]

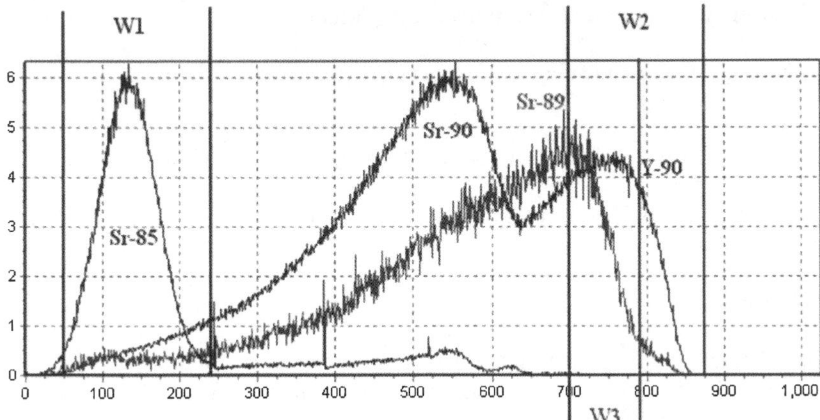

Figure 1 *Visualization of the different regions of interest : W1 centered on the peak of ^{85}Sr, W2 centered on the high energy zone of ^{90}Y peak and W3 possibly containing ^{89}Sr.*

3.1 Instrumentation

The strontium preconcentration by coprecipitation of calcium and strontium phosphates requires magnetic stirrers with heating plate and a centrifuge collecting the precipitate. A drying cabinet is used to gently dry the precipitate of calcium and strontium phosphates. 1220 Quantulus counter is routinely used for the determination of low-level radioactivities.

The extraction of strontium is performed on 2 mL Eichrom™ Sr-spec columns.
Samples are mixed with scintillating cocktail Optiphase Hisafe 3 in 20 mL polyethylene vials from PerkinElmer™.

3.2 Strontium preconcentration by coprecipitation of calcium phosphate

To a volume of 0.5 L acidified water, 4 mL of H_3PO_4 85%, 0.2 mL of a saturated solution of $Ca(NO_3)_2$ and 0.2 mg stable strontium carrier are added in order to achieve a coprecipitation. An activity of about (15±5) Bq of ^{85}Sr tracer is added to the sample.

The sample is mixed and heated to 80-90 °C on a magnetic stirrer. Then precipitation of calcium and strontium phosphates in alkaline medium (pH = 8 – 9) is performed by adding concentrated ammonium hydroxide. Precipitate is allowed to settle for at least 2 hours to one night. Then the supernatant is discarded. The precipitate is collected in a centrifuge tube and rinsed with alkalized water. Precipitate is dried in a warm atmosphere at 80°C.

3.3 Strontium separation using Eichrom™ Sr-spec column

The procedure globally follows the one proposed by Eichrom™ inc. for the determination of ^{90}Sr in water.[9] The elution of strontium is performed by passing 6.0 mL 0.05 M HNO_3 through the column. Ideally, the eluate is conserved to reach the secular equilibrium between ^{90}Sr and ^{90}Y. If not, a correction factor is applied for the growth of ^{90}Y. Then 5 mL of eluate are mixed with 15 mL Optiphase Hisafe 3 in a PE vial. This one is set for counting.

3.4 Counting procedure and ^{90}Sr activity calculation

The same as the Empore membrane. The counting time is fixed to 20 hours.

4 VALIDATION TESTS OF THE EMPORE™ SR RAD DISKS METHOD

The accreditation of this method following the standard NBN EN ISO/IEC 17025 needs to meet some criteria.[8]

4.1 Specifity

The membranes are designed to retain strontium specifically. The disk can retain 3 mg strontium without significant loss (against 8 mg for the Eichrom's 2 mL Sr-spec columns).[10,11] Some information provided by the manufacturer and other users indicate possible interferences. Presence of cations such as Na^+, K^+, NH_4^+ and Ca^{2+} can interfere with strontium fixing.[7,12] If necessary, a preconcentration with the Ca-phosphate coprecipitation technique can be performed to eliminate some interfering elements.

Fixing of other elements like barium, radium and lead, is also observed. It must be noticed that ^{133}Ba, ^{226}Ra and ^{210}Pb could interfere. Absence of ^{133}Ba can be checked by gamma spectrometry. It can be eliminated by precipitating in the form of a chromate at pH 5.5.[13] If radium and strontium are present simultaneously, it is possible to superpose an Empore™ Radium Rad Disk on the membrane for strontium to retain radium. However, routine clearly shows that these interferences are not significantly present in the analysed effluents.

4.2 Uncertainty budget

Type A uncertainties (σ_A), determined by the counting statistics $\varepsilon_{A_{Sr-90}}$, and type B uncertainties, based on a scientific judgment are clearly distinguished.[14] Type B uncertainty (σ_B) takes into account the uncertainty on the activity of the calibration sources certificates and the uncertainty on the volumetry employed during the preparation of solutions (5.55 %). The total relative uncertainty ($\sigma_{A_{Sr-90}}$) spread to 2σ (in %) of the activity in ^{90}Sr is defined in the following general form :

$$\sigma_{A_{Sr-90}} = \sigma_A + \sigma_B = 2 * \varepsilon_{A_{Sr-90}} * 100 + 5.55 \quad (5)$$

The total relative uncertainties met during the analyses for validation are of about a 10 % for aqueous type samples.

4.3 Detection limit and lower limit of detection

The calculation of the detection limit is based on the theory developed by Currie concerning the errors α and β.[15] A value of 0.05 is assigned to α and β which indicates that the conclusion will be exact in 95 % of the cases. Currie indicates that in the case of a well known blank, the detection limit L_d' (in CPM) could be established as follows :

$$L'_d = 3{,}29 * \sigma_{blk} = 3{,}29 * \sqrt{\frac{R_{blk-2}}{t}} \quad (6)$$

where σ_{blk} is the uncertainty on the count rate of the blank in the counting window of ^{90}Y (R_{blk-2}). However, this equation doesn't take yet into account the contribution of ^{85}Sr in W2. It will be associated with the 2σ-uncertainty due to this one :

$$\sigma_{Sr-85} = 2*\sqrt{\frac{R_{85-2} - R_{blk-2}}{t}} \tag{7}$$

Then, the detection limit (L_d) is calculated as follows :

$$L_d = L'_d + \sigma_{Sr-85} \tag{8}$$

The <u>lower limit of detection</u> (LLD) is calculated as follows :

$$LLD = \frac{L_d}{\eta_{90-2} * \eta_{ch} * 60 * V} \tag{9}$$

For a sample volume of 0.5 L, a counting efficiency η_{90-2} of 49 %, and a chemical yield around 80 %, the calculated LLD is approximately 0.02 Bq/L (k=2). This value is definitely lower than the maximum concentration of ^{90}Sr authorized in Belgium in the rejections of aqueous radioactive waste (max. 36 Bq/L).[16]

4.4 Selectivity

To guarantee a good <u>*selectivity*</u> of the technique, pure β-emitter ^{89}Sr must be distinguished from $^{90}Sr/^{90}Y$. Its maximum energy of emission (E_β = 1495.1 keV) is ranging between both: E_β = 545.9 keV for ^{90}Sr and $E_{\beta\ max}$ = 2279.8 keV for ^{90}Y (see Figure 1). Our samples don't contain ^{89}Sr anymore because of its relatively short period ($t_{1/2}$ = 50.57d)[17] and the time of storage before rejection and analysis. In order to verify this, a simple test is carried out. It consists in calculation of a ratio between counting rate in W2 and counting rate in W3. This new ROI is fixed so as to contain 2/3 of the count rate of W2 in the standard $^{90}Sr/^{90}Y$. These windows have the same lower limit. If the ratio exceeds 2/3, it can be due to ^{89}Sr in W3. This test acts just like a warning announcing that the calculated ^{90}Sr activity could probably be over-estimated because of presence of ^{89}Sr.

4.5 Robustness

<u>*Robustness*</u> of the method is ensured by the use of ^{85}Sr tracer allowing the determination of the chemical yield for each analysis. The chemical yield for routine samples is between 60 to 80 %.

4.6 Repeatability

Six aliquots of the same sample were prepared by the same operator and measured in similar conditions. The comparison of the net counting rates R_2 in W2 (^{90}Sr) of the six measurements shows a good repeatability of the method (see table 1). The standard deviation of the set (s) is similar to the counting uncertainty of each measurement (σ_{R_2}).

4.7 Accuracy

No aqueous reference material was available in the laboratory during the validation tests. This test was thus performed on biological reference materials : milk powder (IAEA-152)[18] and dried clover (IAEA-156).[19] These samples had to undergo a thermal mineralization at 500 °C. The salts were recovered in 1 M hydrochloric acid. Then alcalino-terrous elements were preconcentrated by coprecipitation in the form of calcium phosphate in basic medium to get rid of some undesirable elements. In order to eliminate a possible presence of ^{226}Ra an Empore^{TM} Radium Rad Disk was superposed on the usual Sr Rad Disk.
Table 2 presents referenced activities and our results (IPH value). They show a good agreement, in spite of difference in kind of matrix compared to the routine.

4.8 Linearity

In order to test <u>linearity</u> on a relatively large range of ^{90}Sr activity, a series of six deionized waters samples were spiked with increasing activities from 0 to 18.5 Bq/L. Results show an extremely good correlation.

Table 1 *Comparison of the counting rates R_2 in W2 (^{90}Y) for six aliquots of the same sample. Blank contribution in W2 estimated at 0.43 CPM. Counting uncertainty on the Net R_2 (σ_{R_2}) at k=2 ; Standard deviation of the set (s) in CPM.*

Sample n^r	Gross R_2 (CPM)	Net R_2 (CPM)	σ_{R2} (k=2) (CPM)
03/79 a	19.6	19.1	0.37
03/79 b	18.7	18.2	0.36
03/79 c	19.3	18.8	0.37
03/79 d	19.6	19.2	0.37
03/79 e	19.6	19.2	0.37
03/79 f	20.1	19.6	0.37
Mean (CPM)	19.5	19.0	
s (CPM)		0.42	

Table 2 *Reference materials measured to test accuracy of the Empore^{TM}Sr Rad Disks method.*

Reference	Matrix	IAEA value (Bq/kg)	IAEA conf. interval (Bq/kg)	IPH value (Bq/kg)	IPH conf. interv. (k=2) (Bq/kg)
IAEA-152	Milk powder	7.7	7.0 – 8.3	7.9	7.0 – 8.7
IAEA-156	Clover	14.8	13.4 – 16.3	14.0	12.5 – 16.3

4.9 Intra-reproducibilty

This test compares results obtained by two distinct operators within the same laboratory and using the two techniques as described above. 26 radioactive effluents were simultaneously treated and analyzed in parallel.

Figure 2 shows the good correlation between the obtained results. The uncertainties intercept the regression line and the linear coefficient of regression was acceptable. It confirms linearity.

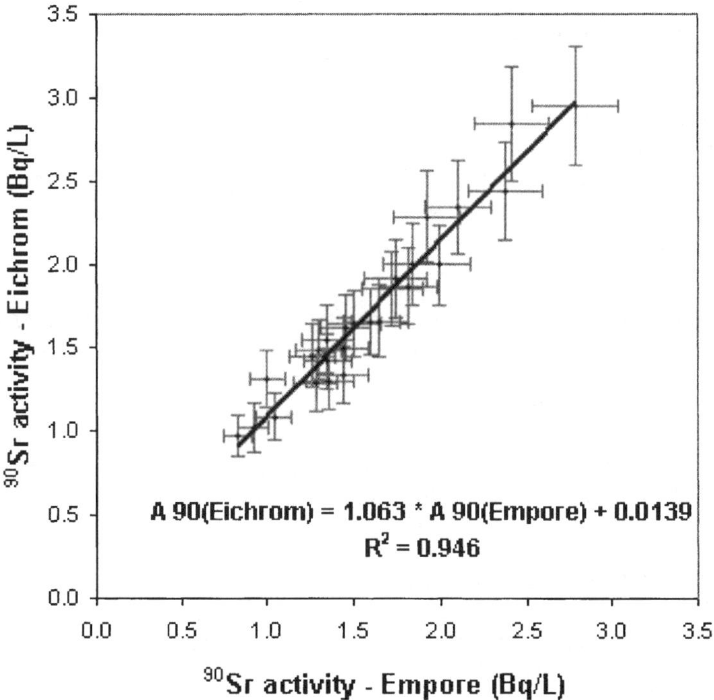

Figure 2 *Visualization of the strontium-90 activities in 26 routine samples of radioactive aqueous waste obtained in parallel by the Empore Rad Disks and Sr-spec columns techniques (enlarged uncertainty k=2).*

4.10 Inter-reproducibility

The laboratory participated to an intercomparison organized by the National Physical Laboratory (N.P.L., U.K.) bearing on an aqueous sample : a mixture of α and β emitters in 2 M HNO_3 medium. A known mass was passed through two superposed membranes: radium disk above the strontium disk.

The official result of sample ABH/03 is 9.631 ± 0.018 Bq/g (at k=1). Our value is 8.96 ± 0.54 Bq/g (k=1). An "u-statistic" test shows an agreement between the two values. IPH *u* value (1.24) is lower than 1.64 and indicates that there is no significant difference between the two results.[20] We conclude to a good inter-reproducibility.

5 CONCLUSIONS

The accreditation in accordance with Belgian standard NBN EN ISO/IEC 17025 is now obtained for the determination of the activity of ^{90}Sr in radioactive wastes by liquid scintillation. It's based on the conclusive results of the different tests carried out to validate the procedure.

The Empore^{TM} Rad disk method allows to save time since it is no more necessary to proceed to a tedious preconcentration in routine. An acidification of water is sufficient before filtration of the sample through the membrane. The strontium fixed on the membrane is recovered by elution using a solution of 0.025 M EDTA in a basic medium.

^{89}Sr could be present with the ^{90}Sr in the effluents at the time of earlier or accidental rejections. Its presence could induce the over-estimation of the ^{90}Sr activity. In order to replace the simple test of presence of ^{89}Sr, it is planned to develop its determination on the basis of this method.

The results of the two biological reference materials indicate that this type of samples can be eventually analysed with the Rad Disk membranes.

References

1 S.C. Goheen, in *DOE Methods for Evaluating Environmental and Waste Management Samples*, ed. Battelle Press, US Department of Energy, Columbus, OH, 1997, Method RP501.

2 S. Scarpitta, J. Odin-McCabe, R. Gasschott, A. Meier and E. Klug, *Health Phys.*, 1999, **76(6)**, 644.

3 U. P. M. Senaratne, W. A. Jester and C. D. Bleistein, *Health Phys.*, 1997, **73(4)**, 601.

4 S.K. Fiskum, R.G. Riley, C.J. Thompson, *J. Radioanal. Nucl. Chem.*, 2000, **245(2)**, 261.

5 S.C. Goheen, in *DOE Methods for Evaluating Environmental and Waste Management Samples*, ed. Battelle Press, US Department of Energy, Columbus, OH, 1997, Method RP515.

6 J. S. Bradshaw and R.M. Izatt, *Acc. Chem. Res.*, 1997, **30**, 338.

7 L.L. Smith, K.A. Orlandini, J.S. Alvarado, K.M. Hoffmann, D.C. Seely and R.T. Shannnon, *Radiochim. Acta*, 1996, **73**, 165.

8 *General requirements for the competence of testing and calibration laboratories.* NBN EN ISO/IEC 17025, 2nd ed. 2005.

9 Analytical procedure SRW01 Rev.1.4. *Strontium 89, 90 in water.* Eichrom Technologies, Inc., Illinois 2003.

10 D. C. Seely, J. A. Osterheim, *J. Radioanal. Nucl. Chem.*, 1998, **236(1-2)**, 175.

11 www.eichrom.com/products/info/sr_resin.cfm

12 E. P. Horwitz, R. Chiarizia and M. L.Dietz, *Solvent Extr. Ion Exch.*, 1992, **10(2)**, 313.

13 *Multi-Agency Radiological Laboratory Analytical Protocols Manual (MARLAP)*, US EPA, US DOE, US FDA, USGS and NIST, 2004, volume II: chapter 14, p.14-58.

14 *Guide to the expression of uncertainty in measurement,* BIPM, IEC, IFCC, ISO, IUPAC, IUPAP, OIML, Switzerland, 1995, ch. 2, p. 3

15 L.A. Currie, *Anal. Chem.*, 1968, **40(3)**, 586.

16 *Arrêté royal du 20.07.2001 portant règlement général de la protection de la population, des travailleurs et de l'environnement contre le danger des rayonnements ionisants*, Moniteur Belge, Brussels, 30/08/2001, p. 29130.

17 *Recommended Decay Data. Decay Data Evaluation Project*, Laboratoire National Henry Becquerel, www.nucleide.org

18 *Report on the intercomparison run IAEA-152 Radionuclides in milk powder*, IAEA/AL/009, Analytical quality control services, IAEA, Vienna, 1988, p.1.
19 *Report on the intercomparison run IAEA-156 Radionuclides in clover*, IAEA/AL/035, Analytical quality control services, IAEA, Vienna, 1991. p. 1.
20 A. Arinc, D.H. Woods, S.M. Jerome, S.M. Collins and coll., in *Environmental radioactivity comparison exercice 2003 (overseas report)*, NPL Report DQL-RN002, National Physical Laboratory, United Kingdom, 2004, p. 25.

Performance of a Portable, Electromechanically-Cooled HPGe Detector for Site Characterization

Ronald M. Keyser and Richard C. Hagenauer
ORTEC, 801 South Illinois Avenue, Oak Ridge, TN 37831

1 Introduction

High-resolution, germanium detectors (HPGe) have long been used for radionuclide characterization of nuclear sites, both buildings and large areas, for the purpose of cleaning up the site. Several systems consisting of a detector, mobile support, collimators and software for data collection and spectrum analysis have been made by national laboratories and commercial companies. The efficacy of these systems is well accepted. The use of HPGe detectors is complicated by the need for cryogenic cooling of the HPGe detector. The traditional method of cooling used liquid nitrogen and the handling of liquid nitrogen in field situations is always difficult. Laboratory type electromechanical coolers have been used for years, but only recently have developments in low-power electromechanical cooling for HPGe detectors made possible the construction of low weight, portable HPGe spectrometers with sufficient efficiency to perform the needed measurements in reasonable count times. The use of a battery-operated cooler simplifies the field operation by eliminating the need for transportation and storage of liquid nitrogen. The 12 volt DC power needed can be easily provided by AC adapter, additional battery packs, or automobile power.

For this work, the liquid nitrogen HPGe detector and MCA in a standard system for field measurements (Iso-Cart) was replaced by a portable HPGe detector with built-in MCA (Trans-SPEC-100). The detector is nominally 40% relative efficiency; a size commonly used for field measurements. The tungsten and steel collimators are cylindrical sleeves surrounding the crystal and projecting forward. The absorbing material is 4.7 mm thick.

The results show this detector performs well in this application and can be used to obtain reasonable detection limits in 1 hour counting times for both field (soil) and buildings (concrete). Details on the angular symmetry and efficiency as a

Figure 1 Trans-SPEC-100 on Iso-Cart.

function of energy and distance are presented both with and without collimation.

2 Experimental

The system is shown in Fig. 1. The detector can be tilted at any angle for the best geometry for data collection, such as for walls or floors. It is not necessary to use low-background detectors because the natural activity in the field of view (FOV) is higher than the activity in the endcap material.

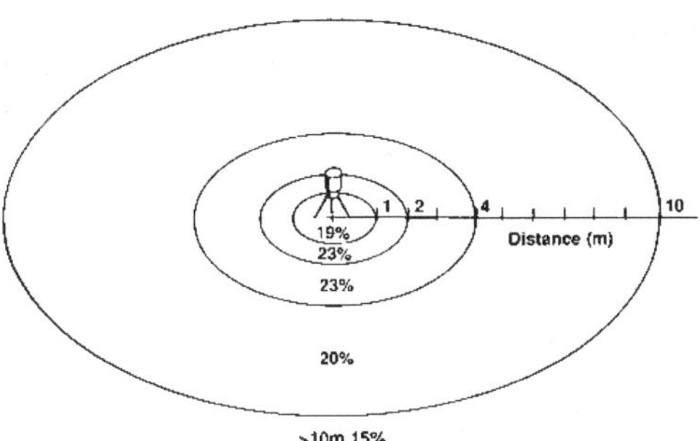

Figure 2 FOV for 1-meter high detector.

The field of view for large, flat surfaces for an uncollimated detector 1 meter above the ground was divided into zones by Beck[1]. The zones are shown in Fig. 2, with the percentages denoting the contribution to the spectrum from the ring of activity. Beck's work was extended to large HPGe detectors by Miller and Shebell[2]. The activity is assumed to be uniformly distributed. Note that more than 80% of the content of the spectrum comes from activity which is outside of the 1-meter diameter circle directly under the detector. The ability of the detector to collect the gamma rays from this area relies on the side efficiency of the detector. For measurements of small areas or areas surrounded by other materials, collimators are used to limit the FOV to the region of interest. Common paving materials have sufficient natural activity to reduce the ability to locate small spills by increasing the background.

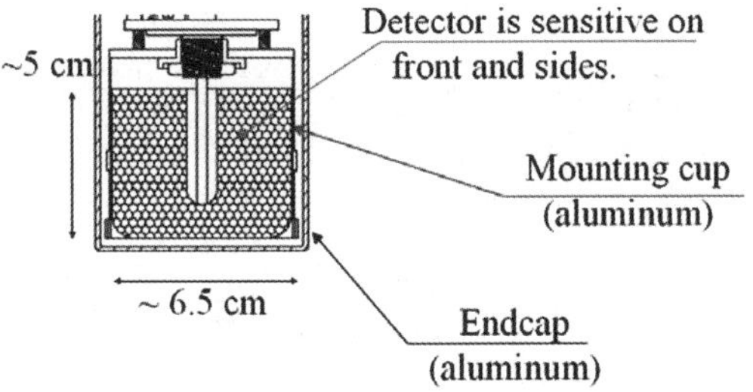

Figure 3 Cross Section of HPGe Detector.

The Trans-SPEC-100 detector efficiency is nominally 40% when measured according to IEEE 325, but this is for a point source at 25 cm in front of the detector. The Trans-SPEC-100 detector is nominally 6.5 cm diameter and 5 cm long. The detector will detect gamma rays on the front and sides, with some decrease at low energies on the sides due to the mounting cup (see Fig. 3), especially for a rugged detector designed to meet ANSI N42.34. Two

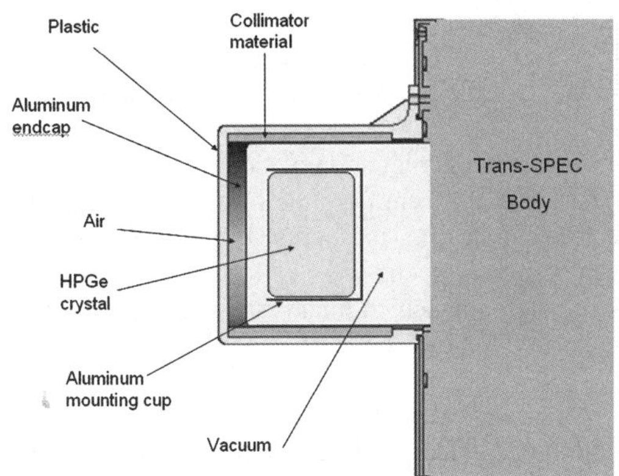

Figure 4 Detector cross section showing collimator position.

different detectors were used for these measurements.

Figure 4 shows the detector with collimator. The metal part of the collimator is cast into the plastic mount. The plastic has some attenuation at low energies. The figure is nominally to scale. The collimator is 85 mm long and extends approximately 18 mm both forward and backward from the ends of the crystal. The collimator is 4.7 mm thick. Not shown are the mounting materials to the rear (toward the Trans-SPEC-100 body) of the crystal which provide additional shielding from that direction.

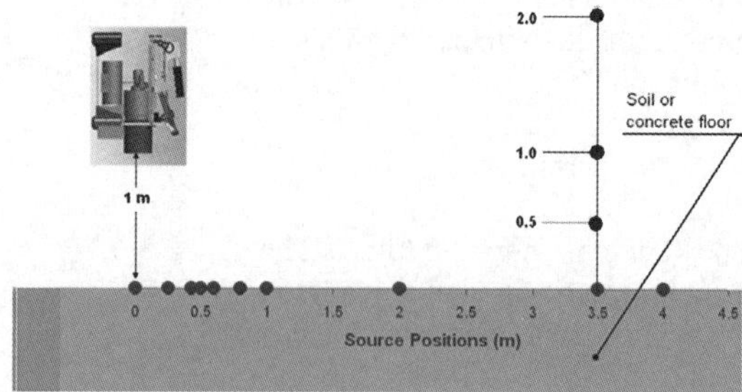

Figure 5 Detector - Source Geometry.

The efficiency at different distances from the center point directly below the detector was measured in the geometries shown in Fig. 5. A multi-nuclide, NIST-traceable point source containing ^{241}Am, ^{109}Cd, ^{57}Co, ^{139}Ce, ^{113}Sn, ^{137}Cs, ^{60}Co, and ^{88}Y was placed at the positions shown, both horizontally and vertically at 3.5 meters distance. Spectra were recorded at each location with sufficient counts to minimize the impact of counting uncertainty.

In addition, the cylindrical symmetry was measured with the detector at 30 cm from the source and the same source placed 30 cm from the center horizontally and at 0°, 90°, 180°, and 270°. This geometry uses both the front face and the side area of the detector.

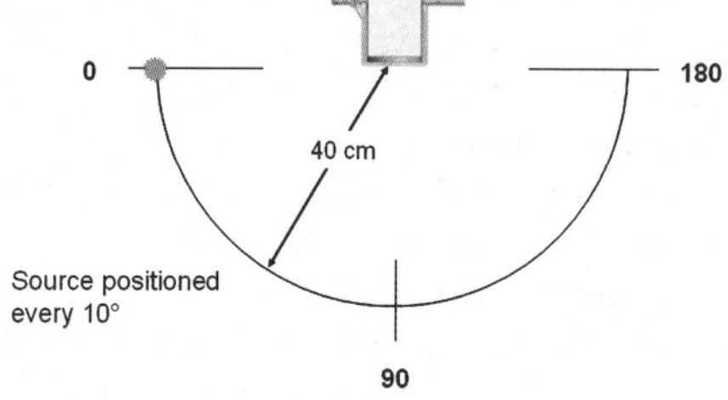

Figure 6 Source positions around detector.

The efficiency at all points was calculated using GammaVision and the polynomial fit to the efficiency points. Peak areas were calculated by the summation method. Both 1461 keV (^{40}K) and 2614 keV (^{208}Tl) were assumed to be constant for all spectra and recorded to ensure quality of the spectra. The MDA values were calculated using the NuReg 4.16 formula.

To measure the reduction of side-incident gamma rays using collimators, the source was positioned at 40 cm from the front of the collimator center point (about 41 cm from the detector endcap). The source (^{133}Ba or ^{60}Co) was positioned at 10° intervals around the center point. Spectra were collected for 1200 seconds in each point for the uncollimated, steel, and tungsten collimators. The peak areas for 81 (a doublet), 356, and 1332 keV were measured for each condition.

The 1-meter calculations are based on calculated efficiencies. For close geometries and collimated detectors the assumptions for this efficiency are not valid so the efficiency in activity calculations should be replaced with a measured efficiency. A geometry that represents the geometry of a small spill or other close-in counting is the 10 cm diameter filter paper source positioned 10 cm from the endcap front. Two different mixed gamma ray emitting sources were used. The efficiencies were calculated using GammaVision and the polynomial fit.

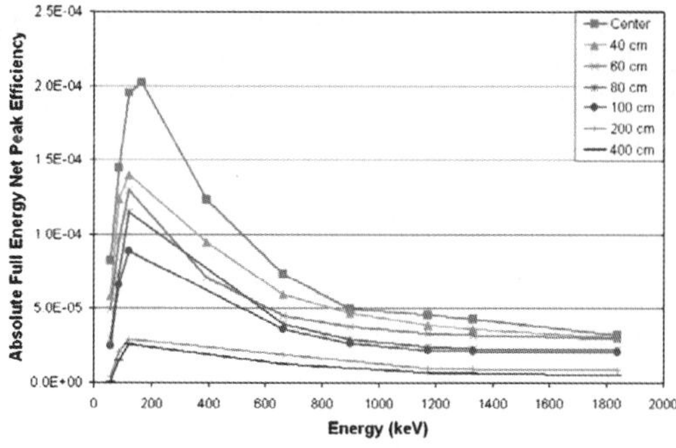

Figure 7 Efficiency for Each Offset by Energy.

3 Results

The point source efficiency at each distance from the center point (as shown in Fig. 5) vs energy is shown in Fig. 7. The dependence on energy is typical for p-type detectors. Note that this is for a point source at the specified distance and is not weighted by the surface area shown in Fig. 2.

The efficiency can also be expressed as relative net peak count rate, where the count rate is normalized to the count rate at the center (or 0 distance). The plot of relative count rate at each energy vs distance is shown in Fig. 8. The reduction of efficiency with distance is somewhat expected. However, the lower energies are reduced more rapidly than the higher energies. This is due to absorption in the air. The 1460 and 2614 keV lines are also plotted. They are nominally constant as expected.

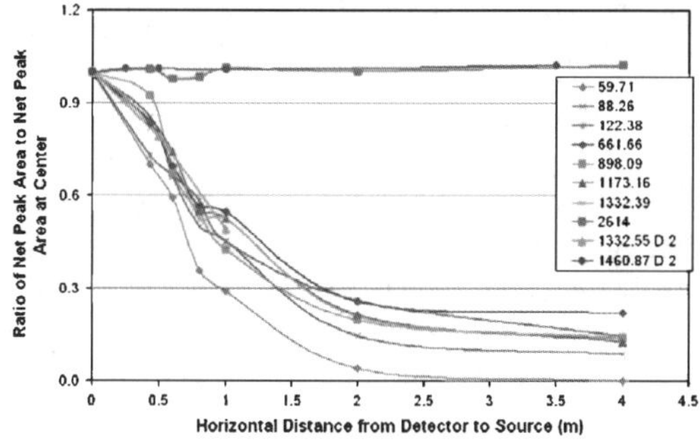

The net peak count rates of the higher energy gamma rays in Fig. 8 do not follow the inverse square dependence. It was suggested[3] that this was due to scattering from the ground. To test this, the source was counted in the vertical positions at 3.5 m distance as shown in Fig. 5. Figure 9 shows the relative net peak count rate as a function of source height. The reduction in count rate with increasing height would indicate the presence of scattering from the ground.

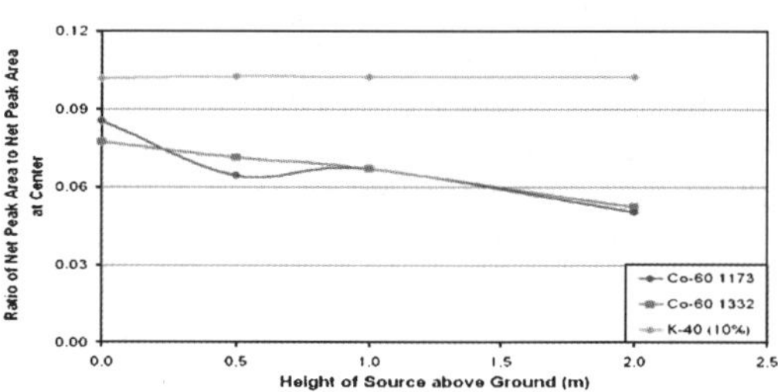

Figure 9 Relative Net Peak Area vs Vertical Height at 3.5 m Distance.

Figure 10 shows the angular response of the detector. The 1461 and 2614 keV peak areas are relatively uniform as expected, but the source peak areas show a reduction in the 90 and 180 directions. This reduction is about 15% and can not be explained from the detector construction. The average data collection time was 7 hours.

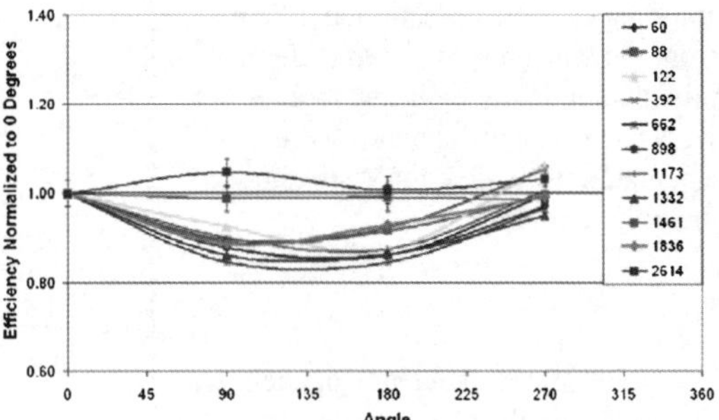

Figure 10 Relative Efficiency by Angle.

The collimator impact for 81 keV is shown in Fig. 11. The counts are normalized to the counts at 90°. The 81 keV lines are almost completely absorbed by either the steel or tungsten. The upturn in the uncollimated curve at 0° and 180° is due to the difference in absorption by the endcap, cup and germanium dead layer for the 0° angle of incidence and the 10° angle of incidence. The combined circular (front face) and cylindrical surface area of the detector perpendicular to the gamma ray flux has a maximum at about 45°, but this is reduced by the change in attenuation by the endcap and germanium dead layer with angle.

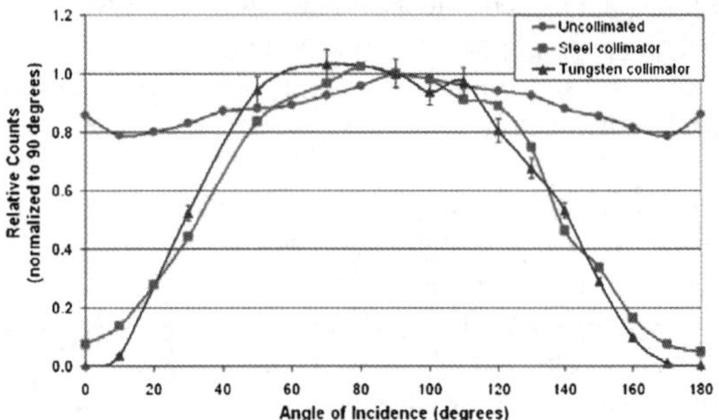

Figure 11 Relative intensity of 81 keV peak vs angle of incidence for two different collimators and uncollimated.

Of more relevance is the 356 keV attenuation as shown in Fig. 12. The figure shows that the steel collimator reduces the side flux by 40% while the tungsten reduced it by more than 80% indicating that the tungsten collimator is more useful in the 100 to 500 keV range than steel.

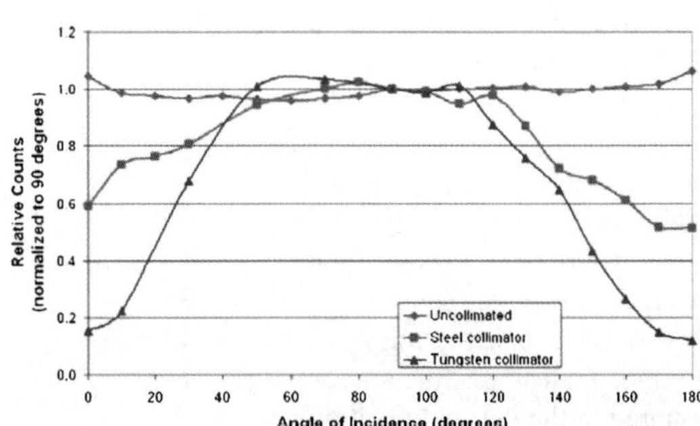

Figure 12 Relative intensity of 356 keV peak vs angle of incidence for two different collimators and uncollimated.

The 1332 keV attenuation is shown in Fig. 13. The tungsten has a higher attenuation than steel (60% vs 70%), but for a significant reduction of 1460 keV (^{40}K) or 2614 keV (^{208}Tl), more tungsten should be used.

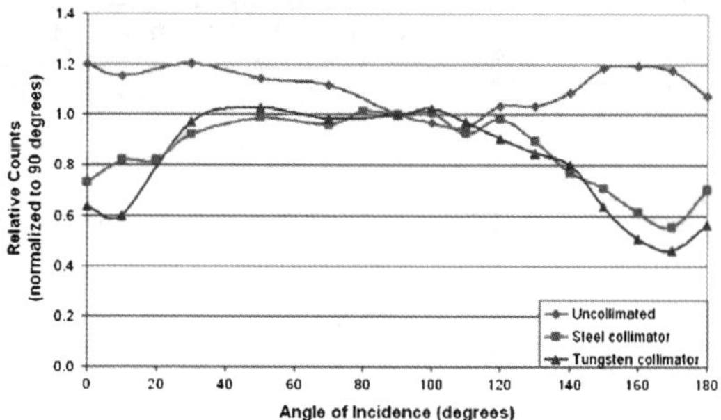

Figure 13 Relative intensity of 1332 keV peak vs angle of incidence for two different collimators and uncollimated.

The absolute efficiency for the filter paper geometry is shown in Fig. 14. The maximum efficiency is about 0.011 at 140 keV. The region from 200 to 400 keV is not well fit because of a lack of un-interfered peaks with sufficient intensity in the region.

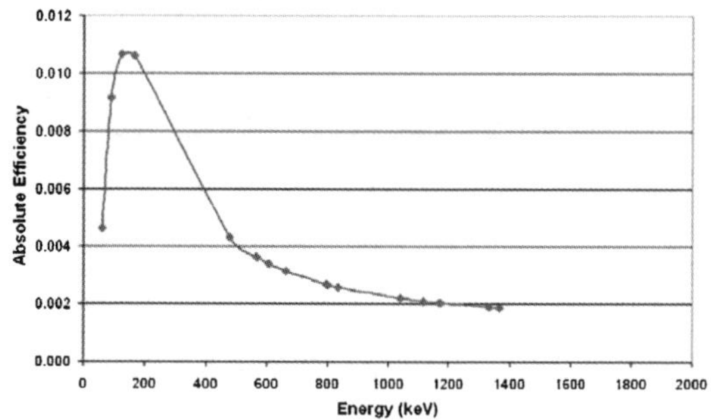

Figure 14 Absolute Efficiency for 10 cm Disk Source at 10 cm Distance from Endcap.

Using the point source efficiencies above, the MDA can be calculated for various nuclides using background counts from this detector. The background spectra from the soil (Tennessee) and the concrete floor of a building used in normal commerce were normalized to a count time of 1 hour and used to determine the MDA values shown in Table 1. The plutonium MDAs are high because of the low yield of gamma rays from ^{239}Pu. The Trans-SPEC-100 detector is 65 mm diameter and 50 mm long. Also shown in Table 1 are the MDA values for a larger detector (80% relative efficiency[4], 72 mm diameter and 87 mm long). As expected, the larger detector has lower MDA values for the same collection time. The ratio of the MDAs (80% detector to Trans-SPEC-100) goes from 0.5 at low energies to 0.12 at high energies. The low-energy efficiency is proportional to the crystal surface area, while the high energy efficiency is proportional to the crystal volume, so these ratios are within expectations. This comparison is for unshielded detectors. Collimating the detectors will reduce the background and for small area measurements (such as spills) will reduce the advantage of the 80% over the 40%. The required detection limits vary with the purpose of the measurement (transportation, waste, or decontamination), country and regulatory agency, but the detection limits for the trans-SPEC-100 satisfy most decontamination (free release) limit requirements.

Table 1. MDA (Bq/m^2) for Concrete Floor and Soil				
		Trans-SPEC-100		80%
Nuclide	Energy (keV)	Concrete Floor	Soil	Soil
Am-241	59.5	42	38	19
U-238	92.6	280	450	140
Pu-239	129.3	1.9E+05	3.9E+05	1.2E+05
U-235	185.7	25	34	9.7
Ba-133	355.0	19	31	6.6
Pu-239	413.7	8.0E+05	1.1E+06	2.7E+05
Cs-134	604.6	15	25	3.5
Cs-137	661.6	11	26	4.8
U-238	1001.1	970	1000	180
Co-60	1332.0	6.8	8.2	1.0

4 Conclusion

These data show that the detector in the Trans-SPEC-100 is suitable for use in field measurements and decommissioning of nuclear sites. The MDAs in the 1-meter mode satisfy most decontamination (free release) limit requirements. Thus, the advantage of using electromechanical cooling over liquid nitrogen cooling does not impose any disadvantage. The use of relatively low-weight collimators for characterizing small areas shows steel collimators can be used for energies below 120 keV, while tungsten should be used to reduce energies up to 800 keV. The measured efficiency for the extended source geometry shows good efficiency for use in counting containers at short distances. The complete system can be used site characterizations and spill measurements, especially in remote areas.

References

1. H. L. Beck, *et al*, In Situ Ge(Li) and NaI(Tl) Gamma-Ray Spectrometry, U.S. Department of Energy, Environmental Measurements Laboratory, HASL-258, September 1972.

2. K. M. Miller and P. Shebell, "Extension of a Generic Ge Detector Calibration Method for In Situ Gamma-Ray Spectrometry" Radioactivity & Radiochemistry, Vol. 11, pp. 14 - 24, December, 2000.

3. Richard Kouzes, Private communication.

4. Relative efficiency compared to 3" x 3" NaI detector as defined in IEEE 325.

NUCLEAR DECOMMISSIONING AUTHORITY RESEARCH AND DEVELOPMENT NEEDS, RISKS AND OPPORTUNITIES

Neil Smart•[1], Andrew Jeapes[2] and Ainsley Francis[2]

[1]NDA, Herdus House, Westlakes Science & Technology Park, Moor Row, Cumbria, CA24 3HU, UK
[2]B170, Nexia Solutions, Sellafield, Cumbria, CA20 1PG, UK

SUMMARY

The NDA remit as set out within the Energy Act includes – "*promote, and where necessary fund, research relevant to nuclear clean up*". The NDA need to underpin delivery and / or accelerate programmes to fulfil the overall mission and technical underpinning of these activities is critical. In this paper we will present consideration of the investment required in Research and Development.

Firstly, NDA set the requirement for nuclear sites to write down within the Life Cycle Baseline Plans (LCBL), at a high level, the proposed technical baseline underpinning the LCBL activities; furthermore we required technology gaps / opportunities in the technical baselines to be outlined in a R&D requirements section to the LCBL. Criteria were established to categorise the R&D in three areas:

- "needs" - those development activities needed to underpin the proposed technical solutions
- "risks" – those activities required to reduce / eliminate key risks to the proposed technical solutions
- "opportunities" – innovations / changes to the technical baselines

The purpose of production of the technical baselines and underpinning R&D requirements is to establish an auditable trail through the LCBL from programme components into how the programme will be delivered. The LCBL 05 was the first programme to attempt this process.

NDA believe the production of the technical baselines and R&D requirements will be of benefit to the SLC's in terms of ensuring a focus on overall programme delivery and not just short term activities. Furthermore, we can ensure that investment in technology is targeted at priority areas, with common issues and requirements identified and solutions on a broader scale will be achievable.

This paper is a short summary of the publication 'NDA – needs, risks and opportunities', available at http://www.nda.gov.uk.

• Corresponding Author – neil.smart@nda.gov.uk

1 INTRODUCTION

The Nuclear Decommissioning Authority (NDA) is a non-departmental public body, set up in April 2005 by the UK Government under the Energy Act 2004 to take strategic responsibility for the UK's nuclear legacy.

The NDA mission is clear: 'To deliver a world class programme of safe, cost-effective, accelerated and environmentally responsible decommissioning of the UK's civil nuclear legacy in an open and transparent manner and with due regard to the socio-economic impacts on our communities'. In line with the mission, the NDA's main objective is to decommission and clean-up the civil public sector nuclear legacy safely, securely, cost effectively and in ways that protect the environment for this and future generations. The NDA does not carry out clean-up work itself but has in place contracts with site licensee companies (SLCs), who are responsible for the day-to-day decommissioning and clean-up activity on each UK site. Individual sites develop LCBLs that set out the short, medium and long-term priorities for the decommissioning and clean-up of each site.

Critical to achieving the NDA main objective and overall mission is to accelerate and deliver clean-up programmes through the application of appropriate and innovative technology. That's why the remit as stipulated in the Energy Act is to: '*promote, and where necessary fund, generic research relevant to nuclear clean-up"*. The NDA have therefore considered the investment required in Research and Development (R&D) both directly and indirectly (i.e. through the Site Licensee Company clean-up programmes) to ensure appropriate delivery of Life Cycle Baseline Plans and to maximise the return on the investment made.

The sites are required to state within the LCBL, at a high level, the proposed technical baseline that underpins the LCBL decommissioning and clean-up activities. In addition the sites are required to identify technology gaps / opportunities in the technical baselines within the R&D requirements section of the LCBL. R&D is categorised in three key areas:

1. The 'needs' - providing solutions to known and common issues
2. The 'risks' - providing options to avoid or mitigate the risks
3. The 'opportunities' - delivering innovative improvements to the LCBL to achieve the NDA's mission

The purpose of including technical baselines and underpinning R&D requirements within the LCBL is to establish an auditable trail through the LCBL and a direct link between the programme components and programme delivery. The LCBLs 05 was the first programme to attempt this process.

Historically, the short-term benefits gained from carrying risks associated with the technical underpinning of projects led to significant cost implications and delays to projects and programmes. Today, NDA believe the technical baselines and identification of R&D requirements will help the SLCs to focus on overall programme delivery and not just short-term activities. In addition, NDA can ensure that investment in technology is targeted at priority areas, with common issues and requirements identified, achieving solutions on a broader scale.

Following the production of the LCBLs 05 and in line with the NDA's mission, NDA have now completed the first review of the 'NDA R&D needs, risks and opportunities'. We have considered the information submitted by the SLCs to the NDA in the LCBLs 05, in terms of the R&D requirements. In doing so, we have compared and contrasted the plans from different sites and evaluated commonalities, differences and potential omissions, with a view to sharing the 'NDA R&D needs, risks and opportunities' across the entire technical supply chain.

2 KEY FINDINGS OF THE LCBL 2005 REVIEW

A top down review of our overall technology needs, risks and opportunities has identified common key issues.

Key issues identified:

2.1 Balance of R&D programmes

Owing to the mature nature of the industry, the vast majority of R&D development activities are integrated directly with on-plant deployment projects and therefore solution driven in their application. Judgements are therefore made about the degree to which remaining knowledge gaps need to be filled or whether the proposed solution is sufficiently robust to deal with the remaining uncertainties.

Given the need to accelerate clean-up programmes in line with the NDA's mission, NDA fully support this approach but recognise the importance of maintaining an appropriate level of underpinning scientific knowledge of the applied processes. In addition NDA will continue to monitor activities to maintain the adequate skills to support the clean-up projects.

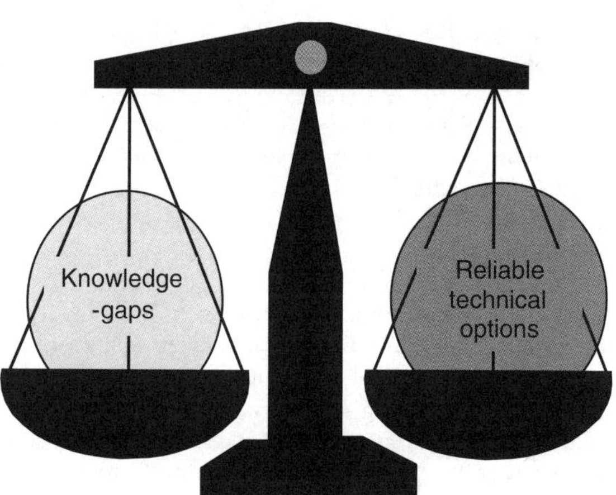

2.2 NDA R&D requirements

The analysis of the full life cycle of existing liabilities overlaid with the need to deliver the NDA's mission yields significant R&D challenges. NDA expect an increase in R&D investment from the SLCs over the next ten years, if delivery is to be assured. We will be monitoring the developing R&D programmes to ensure the activities are being undertaken in line with the delivery of the NDA strategy. We do however expect that confidence in the technology development activities will grow as the programme of clean-up activities accelerates. For this first review, the LCBL plans (quite correctly) did not take account of the proposed acceleration of reactor decommissioning and this is likely to add to the R&D challenges over the next few years.

2.3 Underpinning science

Although most of the R&D activities are focused to support clean-up projects, NDA recognises that these rely upon a strong underlying science base. If clean-up is to be successful it is important to develop the science base as new challenges emerge. To support this, our aim is to stimulate the academic sector to help meet our science needs.

In collaboration with Nexia Solutions, NDA are supporting a series of University Research Alliances (URA) to develop and maintain a network of basic science capability and skills to achieve the short and long-term aims of our mission. Furthermore, the NDA has supported student bursaries, where additional funding has been made available to support PhD projects aligned to the NDA needs, risks and opportunities. Nine awards were made in 06.

The four University Research Alliances are:

- Radiochemistry – University of Manchester
- Particle Science – University of Leeds
- Waste Immobilisation – University of Sheffield
- Materials – University of Manchester

These URA's and a series of smaller University contracts provide a range of underpinning science support to the NDA mission. Also, graduates from PhD programmes are providing an influx of new talent into the decommissioning supply chain.

2.4 Best practice

The LCBL programmes are compiled individually by each site from a 'grass-roots' assessment of the needs of each project, culminating in an overall site plan. This often leads to unique technology solutions bespoke to the project plan for the site. As a result, NDA are encouraging a more integrated approach where sites share proven technology solutions for everyone's benefit and so avoid the cost of bespoke solutions where possible. NDA also fully support the application of proven technology solutions from non-nuclear fields within the nuclear industry.

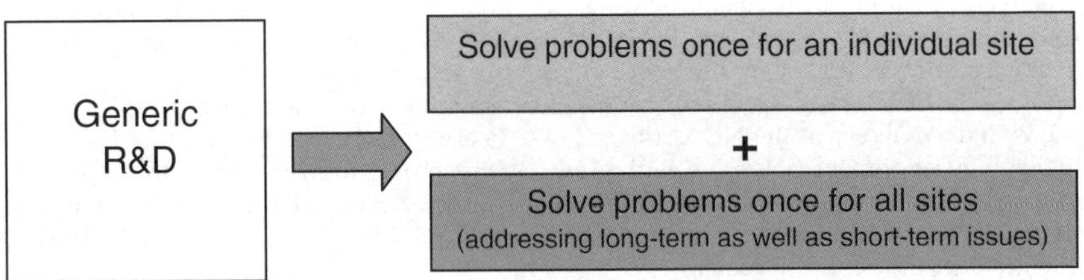

Additionally, international experience in terms of proven technology capability should be considered further. A number of nations have had substantial clean-up programmes over the previous two decades, with proven delivery capability. As improvements in the supply chain management take place within the competitive clean-up market, more proven technology options will be proposed, requiring minimal development activities.

2.5 Contingency activities

A further review of the underpinning research and development activities of high priority projects will establish whether the level of activities (basic science and contingency work) are aligned to the uncertainties and risks associated with the projects. Whilst NDA believe these high priority projects are technically underpinned against current plans, the level of risk associated with the waste streams involved may require more investment to develop contingency options in parallel.

2.5 Safety and environmentally driven research

High standards of safety and environmental performance are a fundamental requirement of the SLCs and NDA are supporting SLC R&D activities that provide improvement in these areas. As the R&D drivers are mainly the same, R&D activities specific to safety and the environment have not been separated out. NDA will actively encourage the SLCs to work together on specific technical projects offering improved safety or environmental performance and therefore mutual benefit.

2.6 Development of common technology solutions

The site LCBLs indicated a range of common problems that would benefit from combining efforts. Examples include a widespread requirement for local and mobile effluent technology and sludge handling technology. Clearly, these processes will have commonality in terms of engineering design, waste disposal, IX technology etc.

2.7 Development of common waste packaging solutions

Most sites need to package and store ILW, whilst awaiting a long-term disposal option. Each waste stream requires process development and assessment for suitability for eventual disposal. There are opportunities to share pre-existing solutions to common waste materials on different sites. There are similar opportunities to establish common formulations for cement and associated fillers, across the waste spectrum, enabling a higher level of confidence in waste packages and security of supply.

2.8 Availability of facilities to meet market needs

Whilst the analysis of 'NDA R&D needs, risks and opportunities' has highlighted work to be carried out, no assessment has been made of the ability of the supply chain to meet these requirements in terms of facilities. One specific area of concern is the changing need in terms of measurement capability and the availability of laboratory facilities and laboratory standards.

2.9 Technology transfer

Accelerating clean-up programmes means accelerating the development process and transfer of new technology options into the market place. A number of interfaces need improvement (a) SLC to SLC (b) Nuclear research organisations to SLC (c) non-nuclear sector to SLC.

2.10 Major technical issues

Major technical needs and opportunities were identified as worthy of highlighting due to importance or widespread interest:

Materials characterisation

The LCBLs identified issues surrounding the characterisation of materials for treatment and disposal.

- Techniques to rapidly assay low levels of radiation and contamination in order to sentence materials for LLW disposal and segregate materials for release as clean or exempt materials
- Development of techniques to characterise contamination of structures
- Development of techniques to characterise site and contaminated land
- Development of techniques to characterise waste properties – radiochemical, chemical and physical, including facilities for ILW characterisation

Waste processing

- Development of proven sludge handling techniques
- Remote handling techniques for fuel debris and highly activated materials
- Methodology and techniques for waste segregation
- Recycle of materials
- Graphite management

Management of strategic nuclear materials

- Development of immobilisation technology for separated plutonium
- Process development for the conversion of uranium hexafluoride to a more stable form
- Long-term options for the UK approach to the management of spent fuel

Plant termination

- Improved decontamination technology to either enable man access or waste re-categorisation
- Improved effluent management to process decontamination reagents
- Development of local and mobile effluent treatment capability
- Technology to carry out size reduction of large items
- Remote dismantling technologies

Site restoration

- Surveying and characterisation of land contamination
- Ground remediation technology for active and non-active contaminated land
- Development of consistent protocols to underpin site end point considerations for a wide variety of sites

The quality of the 'NDA R&D needs, risks and opportunities' submissions by the SLCs varied considerably. We will be working to ensure that future submissions are of a consistent and high quality.

3 CONCLUSIONS

The NDA are committed to the identification of Research and Development Needs, Risks and Opportunities underpinning the NDA mission. These are published in the open forum, where ever possible, with a view to the encouragement of supply chain engagement and development of an innovation driven culture.

THE PERFORMANCE OF UK AND OVERSEAS LABORATORIES IN PROFICIENCY TESTS FOR THE MEASUREMENT OF ^{241}Am

A.V. Harms, J.C.J. Dean, C.R.D. Gilligan and S.M. Jerome

National Physical Laboratory, Hampton Road, Teddington TW11 0LW, UK.
E-mail: arvic.harms@npl.co.uk

1 INTRODUCTION

Since the late 1980s, the National Physical Laboratory (NPL) has organised 12 proficiency test exercises for a range of radionuclides including fission products and activation products of both reactor construction materials and nuclear fuel.[1,2] This has, over the years, grown to encompass a number of actinides including ^{241}Am. This radionuclide, which has a half-life of 432.6(6) y and has both alpha and gamma emissions,[3] is of environmental concern as it is directly released to the environment via fallout and from fuel reprocessing activities.[4,5] The activity level of ^{241}Am in the environment will increase due to ingrowth from ^{241}Pu, a low-energy beta emitter with a half-life of 14.290(6) y.[6]

Americium-241 can be measured by employing several different measurement strategies: alpha spectrometry gamma spectrometry, and mass spectrometry. Measurement of ^{241}Am with alpha spectrometry requires radiochemical separation from other radionuclides in the sample (especially ^{238}Pu, whose 5.456 and 5.499 MeV peaks interfere with the 5.443 and 5.486 MeV peaks of ^{241}Am).[7] In general, the participants used radiochemical techniques involving ion-exchange chromatography and (more recently) extraction chromatography. In virtually all cases, ^{243}Am was used as the chemical yield tracer. Measurement of ^{241}Am with gamma spectrometry is relatively straightforward since (i) it does not require any radiochemistry or pre-concentration at the ^{241}Am activity concentration levels provided and (ii) most commercially available multi-nuclide calibration sources contain ^{241}Am. However, care must be taken to correct for attenuation of the 59.5 keV photon if the composition of the exercise sample is different from the calibration source used.[8] Measurement of ^{241}Am with mass spectrometry requires in general chemical separation from other actinides and matrix elements.[9]

The data gathered by NPL to date allow results obtained using these different techniques to be compared and demonstrate how laboratories employing them fared in the proficiency test exercises NPL has conducted.

2 METHOD AND RESULTS

The radionuclide ^{241}Am was included in 10 proficiency test exercises (Table 1). The first exercise that included ^{241}Am offered a relative simple mixture of ^{239}Pu and ^{241}Am. The subsequent 9 exercises offered samples that contained a more complex mixture of alpha and beta emitters. The radionuclides ^{90}Sr, ^{238}Pu, ^{239}Pu and ^{241}Am were always included, while ^{3}H (7 exercises), ^{99}Tc (5 exercises), ^{210}Pb, ^{226}Ra and ^{238}U (4 exercises each), ^{55}Fe and ^{89}Sr (2 exercises each) and ^{35}S, ^{63}Ni and ^{244}Cm (1 exercise each) were included on a less regular basis. Prior to 2000, only one activity level with activity concentrations of the order of ~10 Bq kg^{-1} was available. Since 2000, a second, higher activity level (with activity concentrations of the order of ~2 Bq g^{-1}) has been offered for the alpha / beta emitter sources, while since 2002 a separate low-energy beta emitter mixture containing ^{3}H and at least one other low-energy beta emitter, such as ^{14}C, ^{35}S or ^{129}I, has been offered.

Table 1 *NPL Environmental Radioactivity Proficiency Test Exercises*

NPL Exercise	*Activity concentration* ^{241}*Am* (Bq kg^{-1})	^{241}*Am results alpha spec*	^{241}*Am results gamma spec*	*Other radionuclides present in the alpha/beta sources*
3 (1992)	8.6(1) – 10.4(1)	11	2	^{239}Pu
4 (1993)	11.30(5) – 14.50(5)	14	7	^{3}H, ^{90}Sr, ^{99}Tc, ^{238}Pu and ^{239}Pu
5 (1995)	16.6(8) – 29.5(15)	12	11	^{3}H, ^{90}Sr, ^{99}Tc, ^{238}Pu and ^{239}Pu
6 (1996)	10.13(9) – 16.35(10)	11	13	^{3}H, ^{35}S, ^{90}Sr, ^{238}U, ^{238}Pu and ^{239}Pu
7 (1998)	4.82(7)	7	5	^{3}H, ^{89}Sr, ^{90}Sr, ^{238}U, ^{238}Pu and ^{239}Pu
8 (2000)	2.526(12) and 2035(9)	17	8	^{3}H, ^{55}Fe, ^{63}Ni, ^{90}Sr, ^{238}U, ^{238}Pu, ^{239}Pu and ^{244}Cm
9 (2001)	2.58(6) and 2060(18)	10	6	^{3}H, ^{90}Sr, ^{99}Tc, ^{210}Pb, ^{226}Ra, ^{238}Pu and ^{239}Pu
10 (2002)	2.47(6) and 1996(18)	19	10	^{90}Sr, ^{210}Pb, ^{226}Ra, ^{238}Pu and ^{239}Pu
11 (2003)	3.058(23) and 2466(18)	27	8	^{90}Sr, ^{99}Tc, ^{238}Pu and ^{239}Pu
12 (2005)	11.99(4) and 3691(13)	24	8	^{55}Fe, ^{89}Sr, ^{90}Sr, ^{238}U, ^{238}Pu and ^{239}Pu

To prepare the mixed alpha/beta sources, standardised single radionuclide solutions were combined and diluted as necessary. This was performed in accordance with established procedures that have been independently accredited by the United Kingdom Accreditation Service (UKAS) for the production of solution standards of radioactivity. The exact composition of the sources varied with each exercise, but in general consisted of a 1 – 2 M nitric acid solution containing 50 ppm Sr and 50 ppm Y as carriers for radioisotopes of these elements.

In total, 235 results for ^{241}Am were returned from all the exercises combined, with 152 results obtained by alpha spectrometry, 78 results obtained by gamma spectrometry, 3 results obtained by an unknown method and 2 results obtained by mass spectrometry. In recent exercises, about a quarter of the ^{241}Am results were obtained by gamma spectrometry, whereas about half of the ^{241}Am results from the exercises conducted in the late nineties had been obtained by this technique. The number of results obtained by mass spectrometry has been negligible and these results were not particularly accurate. In the early exercises, the participants almost exclusively originated from the UK measurement community, while in the later exercises more overseas laboratories took part. In the most recent proficiency test exercise (2005), 42% of the participants represented overseas laboratories. The return for ^{241}Am was generally high (e.g., in the 2005 exercise 32 results were returned from 36 send-out samples, although two participants submitted more than one result by using both alpha and gamma spectrometry).

The returned data was received as an activity concentration with a standard uncertainty and was subsequently converted into a deviation (in %) from the assigned NPL values. This deviation from the assigned values is given by:

$$D = 100\frac{L-N}{N}\% \qquad (1)$$

The standard uncertainty (k=1) of the deviation is given by:

$$u_D = 100\frac{L}{N}\sqrt{\left(\frac{u_L}{L}\right)^2 + \left(\frac{u_N}{N}\right)^2}\,\% \qquad (2)$$

where:

D – deviation from the assigned value (unit: %)
L – laboratory value (unit: Bq kg^{-1})
N – assigned value (unit: Bq kg^{-1})
u_D – standard uncertainty of the deviation (unit: %)
u_L – standard uncertainty of the laboratory value (unit: Bq kg^{-1})
u_N – standard uncertainty of the assigned value (unit: Bq kg^{-1})

Table 2 *Alpha spectrometry and gamma spectrometry results*

	Arithmetic mean (%)	*Weighted mean (%)*	*Median*	*N*	*Range (%)*	*–/+ values (%)*
Alpha spec raw data	–1.6(16)	1.8(16)	–2.3(7)	152	–62 to 148	66 / 34
Gamma spec raw data	5.0(19)	–1.0(8)	1.9(13)	78	–25 to 81	40 / 60
Alpha spec corrected data	–3.2(7)	–3.1(7)	–2.3(6)	142	–29 to 19	68 / 32
Gamma spec corrected data	1.9(12)	–1.2(8)	1.5(14)	74	–25 to 37	42 / 58

Analysis of the raw data (Table 2 and Figures 1 and 2; the bars in the figures represent the standard uncertainty of the deviation) suggested a difference between the results obtained by alpha spectrometry and the results obtained by gamma spectrometry, although this difference was only significant for the medians. Medians have a better statistical "robustness" than mean values and are less affected by outlier effects.[10] It was decided to use a method based on the interquartile range (IQR) to identify possible outliers.[11] Using this method, a value is considered as an outlier if it is either less than the first quartile (25%) minus 3 times the IQR or more than the third quartile (75%) plus 3 times the IQR.

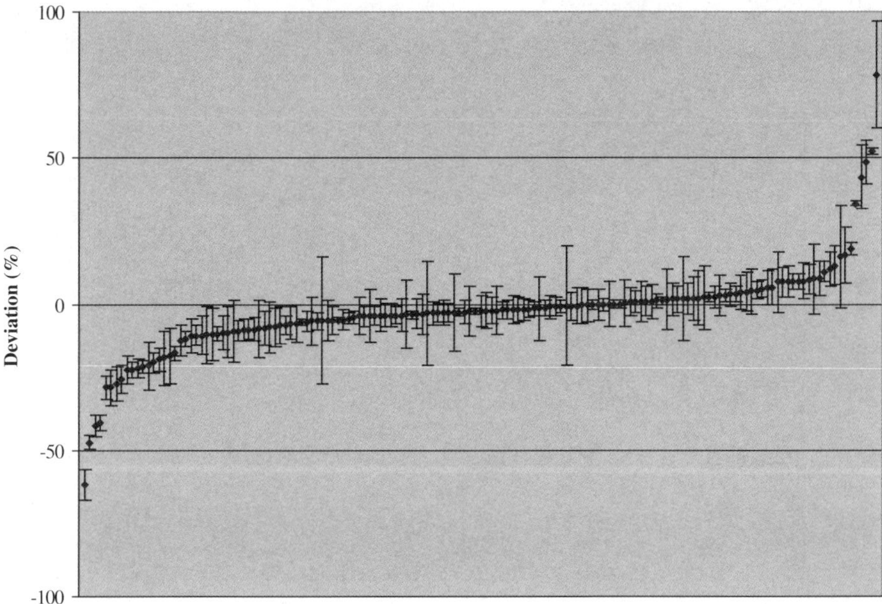

Figure 1 *Alpha spectrometry results; raw data*

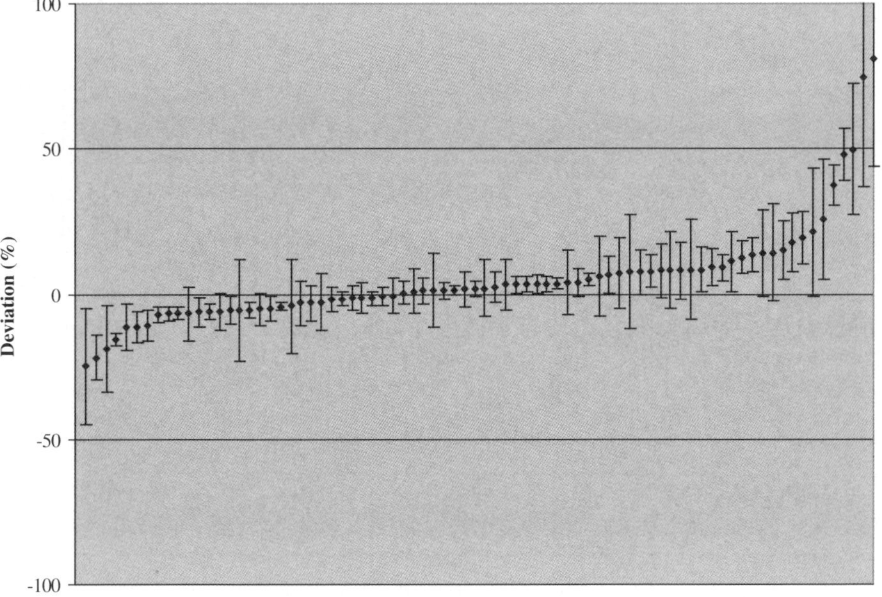

Figure 2 *Gamma spectrometry results; raw data*

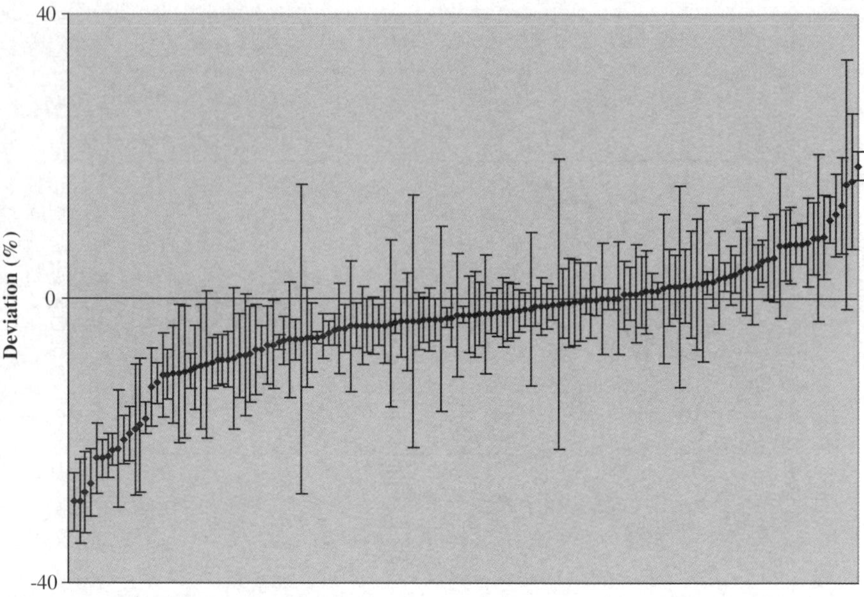

Figure 3 *Alpha spectrometry results; corrected for outliers*

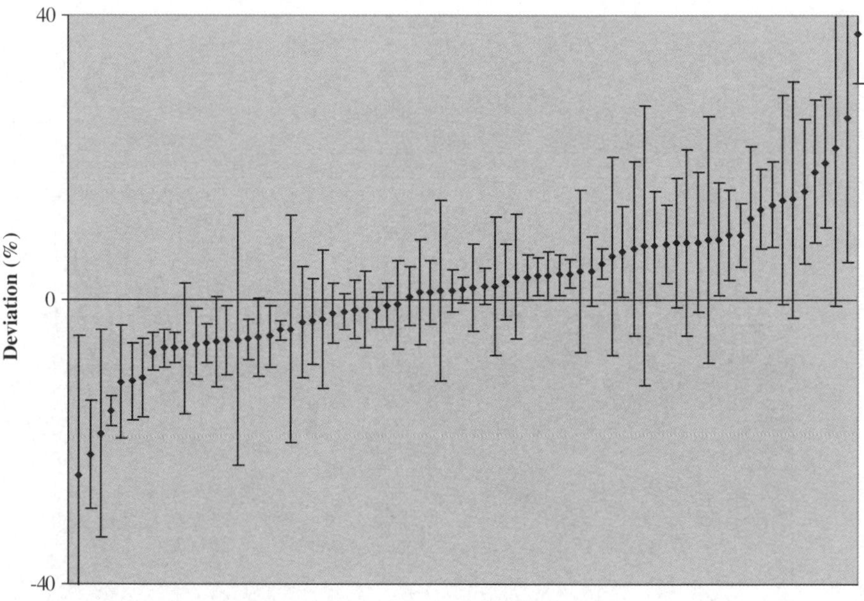

Figure 4 *Gamma spectrometry results; corrected for outliers*

This method identified 14 outliers (Table 2 and Figures 3 and 4) and resulted in a significant difference between the arithmetic means of the alpha spectrometry results and gamma spectrometry results at a 99% confidence level (the t-value being 3.87, which is larger than the critical t-value of 2.61 for 214 degrees of freedom), while the medians of both data sets were significantly different as well. Application of the non-parametric Wilcoxon Rank Sum test,[12] which does not assume a normal distribution and does not require the removal of outliers, also resulted in a significant difference at a 99% confidence level between the alpha spectrometry results and gamma spectrometry results (the absolute z-value being 3.74). More than two thirds of the alpha spectrometry deviations had a negative value, compared to 42% of the gamma spectrometry deviations.

The alpha spectrometry results were also significantly different at a 99% confidence level from the assigned NPL values (which deviations are 0% by definition). Application of the non-parametric Wilcoxon Signed Rank test,[13] which, like the Rank Sum test, does not assume a normal distribution and does not require the removal of outliers, also resulted in a significant difference at a 99% confidence level between the alpha spectrometry results and the assigned NPL values (the absolute z-value being 3.72).

The gamma spectrometry results were not significantly different from the assigned NPL values (Table 2), which was supported by Wilcoxon Signed Rank test (the z-value being 2.24).

3 CONCLUSION

It can be concluded that for ^{241}Am the alpha spectrometry results were significantly different from both the gamma spectrometry results and the assigned NPL values. The arithmetic mean, weighted mean and median value for the alpha spectrometry deviations corrected for outliers were –3.2(7)%, –3.1(7)% and –2.3(6)%, respectively.

It is not easy to explain the significant negative bias of the ^{241}Am alpha spectrometry results; it is unlikely that either incomplete separation from ^{238}Pu (which would yield a positive deviation) or a systematic negative bias of the assigned NPL values (which would also affect the gamma spectrometry results) can provide an explanation. However, problems with the use of the ^{243}Am tracer cannot be ruled out; degradation of the alpha spectrum may result in deformed peaks and tailing which would make it difficult to completely deconvolute the peaks of the two isotopes (which have a relative small energy difference of only ~0.2 MeV).[14] Any loss of ^{241}Am counts in the ^{243}Am peaks will result in a negative bias. The results for underspiked samples (where ^{241}Am/^{243}Am ratio is > 1), prepared with microprecipitation as the source preparation technique and analysed with algorithms unable to correct for ^{241}Am peak tailing are expected to be most affected.

References

1 A. Arinc, D.H. Woods, S.M. Jerome, S.M. Collins, A.K. Pearce, C.R.D. Gilligan, K.V. Chari, M. Baker, N.E. Petrie, A.J. Stroak, H.C. Phillips and A.V. Harms, in *Environmental Radioactivity Comparison Exercise 2003 (Overseas report)*, NPL Report DQL-RN 002 (this report contains references to all earlier Exercises), 2004. http://www.npl.co.uk/ionrad/publications

2 A. Harms, C. Gilligan, A. Arinc, S. Collins, S. Jerome, L. Johansson, D. MacMahon, and A. Pearce, in *Environmental Radioactivity Comparison Exercise 2005*, NPL Report DQL-RN 015, 2006. http://www.npl.co.uk/ionrad/publications

3 DDEP, 2005. The Decay Data Evaluation Project. www.nucleide.org

4 M. Eisenbud and T. Gesell, in *Environmental Radioactivity. From Natural, Industrial, and Military Sources*, 4th Edn., Academic Press, California, USA, 1997.

5 P.E. Warwick, I.W. Croudace and R. Carpenter, *Appl. Rad. Isot.*, 1996, **47**, 627.

6 NuDAT, ENSDF. www.nndc.bnl.gov/nudat2

7 P.E. Warwick, I.W. Croudace and J.S. Oh, *Anal. Chem.*, 2001, **73**, 3410.

8 N.H. Cutshall, I.L. Larsen and C.R. Olsen, *Nucl. Instr. Methods*, 1983, **206**, 309.

9 M. Agarande, S. Benzoubir, P. Bouisset and D. Calmet, *Appl. Rad. Isot.*, 2001, **55**, 161.

10 J.W. Müller, *J. Res. Natl. Inst. Stand. Technol.*, 2000, **105**, 551.

11 W. Mendenhall and T. Sincich, in *Statistics for Engineering and the Sciences*, 4th Edn., Prentice-Hall, New Jersey, USA, 1995, p. 59

12 W. Mendenhall and T. Sincich, in *Statistics for Engineering and the Sciences*, 4th Edn., Prentice-Hall, New Jersey, USA, 1995, p. 928

13 W. Mendenhall and T. Sincich, in *Statistics for Engineering and the Sciences*, 4th Edn., Prentice-Hall, New Jersey, USA, 1995, p. 937

14 Z. Lin, A. Berne, B. Cummings, J.J. Filliben and K.G.W. Inn, *Appl. Rad. Isot.*, 2002, **56**, 57.

CURRENT IAEA ACTIVITIES AND FUTURE PLANS FOR THE ALMERA NETWORK

Chang-Kyu Kim, Paul Martin and Gyula Kis-Benedek

Physics, Chemistry and Instrumentation Laboratory, Agency's Laboratories Seibersdorf, International Atomic Energy Agency, A-1400, Vienna, Austria (e-mail: c.k.kim@iaea.org)

1 INTRODUCTION

With reference to the measurement of artificial and natural radionuclides in environmental samples, activities of the Agency's laboratories in Seibersdorf include the preparation, characterisation, and distribution of reference materials of terrestrial origin, organisation of interlaboratory comparison exercises and proficiency tests, co-ordination of the activities of the ALMERA network, provision of training and expert assistance to scientists from Member States' (MS) laboratories, and provision of analytical and radioanalytical support for the benefit of the programmes and projects of the IAEA and its Member States.

The IAEA's network of Analytical Laboratories for the Measurement of Environmental RAdioactivity (ALMERA) was formally established in 1995 with 53 laboratories from 26 different countries when the IAEA issued a request to its Member States to nominate laboratories for membership. The members of the network are gradually increasing in number. Currently the network comprises 106 laboratories from a total of 67 countries. The nomination of laboratories for membership in ALMERA has to be made by their governments and the IAEA has to be informed about the nominations through the Permanent Missions of the Member States to the Agency. There is no deadline for such a nomination.

The network is a technical collaboration of existing institutions. It provides an operational framework to link expertise and resources, in particular when a boundary-transgressing contamination is expected or when an event is of international significance.

The IAEA helps the ALMERA network of laboratories to maintain their readiness by coordination activities, by development of standardized methods for sample collection and analysis, and by conducting interlaboratory comparison exercises and proficiency tests as a tool for external quality control.

In 1996, the first interlaboratory comparison exercise of the ALMERA network was organised to compare the performance of the network laboratories for the determination of anthropogenic and primordial α-, β- and γ-emitting radionuclides in a soil and sediment material. This first activation of the network was executed as a preliminary test to determine which laboratories could perform the necessary radiochemical analyses in support of the Agency's upcoming Mururoa project.

In 1997, the Agency organised the first ALMERA workshop[1] which was held in Vienna and was attended by Agency staff and representatives of 24 laboratories from 15 countries. The analytical results of the first ALMERA intercomparison exercise were discussed, the role of ALMERA was defined and a strong commitment to continue the work of ALMERA was expressed by all participants. During 1996-97 the ALMERA network supported the IAEA project on the evaluation of the radiological situation at the atolls of Mururoa and Fangataufa. In 2001 an ALMERA proficiency test on ^{238}Pu, $^{239+240}$Pu, ^{241}Am and ^{90}Sr in spiked soil was carried out[2].

Beginning with the 2004/05 biennium, the Agency's Seibersdorf laboratories have started several activities with the aim of making the ALMERA network more active. As ALMERA laboratories are expected to be those which would be active in the event of any urgent action (including emergency) situations, there is a particular need for them to have available the tools for undertaking sampling and analysis on a fast turnaround basis. In addition to activities linked to ALMERA, there is considerable general demand from IAEA Member State laboratories for interlaboratory comparison exercises and the provision of recommended procedures for environmental sampling and analysis. This paper summarises these efforts.

2 RECENT AND CURRENT ACTIVITIES

2.1 The 2nd ALMERA Network Coordination Meeting

The second ALMERA network coordination meeting took place in Trieste (Italy) on 15 November 2005 and was hosted by the International Centre for Theoretical Physics (ICTP).

The overall aim of the meeting was to evaluate the current status of the ALMERA network laboratories and to help to improve their technical competence through harmonization of sampling, monitoring and measurement protocols and staff training. The meeting was also addressed to define the structure of the ALMERA network and future proficiency tests and intercomparison trials to be organized by the IAEA to help the laboratories to maintain and improve the quality of their analytical measurements. 45 participants from 29 different institutions attended the meeting. The participants of the ALMERA coordination meeting recommended the following[3] :

1) An intercomparison exercise or a proficiency test should be organised annually. The following matrices were proposed: soil, water, vegetation and air filter which are of interest for routine monitoring and emergency situations. The analytical techniques that were recommended for inclusion in the proficiency tests are summarised in Table 1.

2) The proficiency testing materials should be prepared by characterization in expert laboratories and the target value and the associated combined uncertainty should be assigned according to ISO Guides 30-35[4].

Table 1 *Analytical techniques recommended for inclusion in proficiency tests*

gamma-spectrometry (HPGe)	routine monitoring, rapid measurements
gamma-spectrometry NaI(Tl)	rapid measurements
gross alpha/gross beta	rapid measurements (liquid samples)
TRU by alpha spectrometry	routine monitoring
^{90}Sr	routine monitoring, rapid measurements
^{3}H	routine monitoring, rapid measurements
natural radionuclides	routine monitoring

In addition to the above reporting times, the laboratories participating in the ALMERA proficiency tests will be also asked to report analytical results to the organiser at the reporting time indicated for the IAEA world-wide open proficiency test (i.e. normal reporting). It should be noted that the approach for evaluation of the results for the rapid and normal reporting time may be carried out on a different basis. This is because it can not be expected that the same accuracy would be reached by a rapid analysis as by a normal analysis.

3) Considering that the ALMERA network was designed to support measurement of radionuclides in issues of international concern or emergency situations, where the measurement results should be ready in a short time, the laboratories participating in the ALMERA proficiency tests should report their analytical results to the organizer according to the time schedule given in Table 2.

Table 2 *Reporting time recommended of the results in proficiency tests*

gamma-spectrometry (HPGe & NaI(Tl))	1 day
gross alpha/gross beta	1 day
transuranic by alpha spectrometry	1 week
^{90}Sr	1 week
^{89}Sr	1 week
^{3}H	3 days

4) The performance evaluation results of the interlaboratory comparison exercises performed in the frame of the ALMERA network are not anonymous for those laboratories nominating to participate as ALMERA members.

If an ALMERA member wants to keep the evaluation result of his/her participation anonymous, he/she will have the option to only take part in the IAEA world-wide open proficiency test. In this case, his/her results will not be included in the ALMERA report. In addition, the statistical approach used to evaluate the analytical results of the ALMERA network proficiency test will be adapted in the future to take the reporting time into consideration.

5) It was recommended that the target values for each proficiency test should be released soon after the deadline for data submission for the world-wide test. It was also requested that other field intercomparison exercises (gamma does rate evaluation by in-situ gamma-spectrometry, sampling in aquatic systems, etc.) should be organized.

6) To facilitate interactions between the ALMERA laboratories, for the period 2007-2009 the network is subdivided into the following three regional groups:

- Africa-Europe-Middle East;
- North and Latin America;
- Asia-Pacific.

The groups were defined by the IAEA according to the geographical distribution of participating laboratories. Each ALMERA laboratory is requested to indicate the group to which it wants to belong. In the case of no indication being given, the IAEA will assign the ALMERA laboratory to a regional group on the basis of its geographical location. If in future a laboratory wishes to join another group for any reason it should be allowed to do so. In case of countries geographically located between different regional groups and having two or more laboratories in ALMERA, it should be allowed for the different laboratories to belong to different regional groups.

It should be emphasized that assignment to one regional group will not preclude a laboratory from taking part in ALMERA activities of another regional group. Nevertheless, in order to simplify organization, each laboratory should formally belong to only one regional group.

Each regional group is coordinated by an ALMERA regional focal point.

7) In order to improve communication between members, it was recommended that an ALMERA network website should be developed so that all members can search for diverse information regarding the activity of the ALMERA network. It should include the following data:

- ALMERA news

- The analytical capabilities of ALMERA member laboratories

- The results of proficiency tests

- A database of sampling method and analytical procedures of radionuclides in environmental samples.

- Minutes of the ALMERA meetings

- A questionnaire on the analytical capabilities of ALMERA member laboratories, research projects being implemented, etc.

8) For the period 2007-2009, the focal point for the Africa-Europe-Middle East region will be the IAEA central ALMERA coordinator, i.e. the Chemistry Unit of the IAEA Physics, Chemistry and Instrumentation Laboratory (PCI).

For the period 2007-2009, the focal point for the Asia-Pacific region will be the Korea Institute of Nuclear Safety (KINS).

The fifth ALMERA coordination meeting will take place at the Brazilian National Commission for Nuclear Energy (CNEN), Instituto de Radioprotecao e Dosimetria, Rio De Janeiro, Brazil, in November 2008.

2.2 ALMERA soil sampling intercomparison exercise

Advances in analytical techniques and improved laboratory practice have reduced many sources of uncertainties which can originate during laboratory analytical procedures. However, the assessment of uncertainties associated with sampling of environmental components has not been fully considered in the past, since collaborative field studies require considerable organisational efforts. Sampling and sample preparation/processing are known to carry large, but typically unknown, uncertainty contributions to the final analytical data and there is a lack of qualitative and quantitative data on the comparability of results achieved by the different sampling methods. ISO/IEC 17025[5] reports that sampling is a factor to be considered as a contributor to the total uncertainty of measurement.

To this end, the IAEA organized a soil sampling intercomparison exercise for selected laboratories of the IAEA ALMERA network. The objective was to compare the soil sampling protocols used by the different participating laboratories, when they were asked to determine the mean value of several radionuclides in an agricultural area of about 10000 square meters. The radionuclides to be considered in planning the sampling exercise were those that require radiochemical separation (^{90}Sr, Pu, ^{241}Am, ^{238}U) and a test portion (quantity of material, of proper size for measurement of the concentration or other property of interest, removed from the test sample) ranging from 10 to 50 g, depending on the activity concentration of the radionuclide[6].

The intercomparison exercise took take place from 14 to 18 November 2005 in an agricultural area qualified as a "reference site" (area, one or more of whose element concentrations are well characterised in terms of spatial and temporal variability[7], in the frame of the SOILSAMP international project, funded and coordinated by the Italian Environmental Protection Agency, APAT, Italy) and aimed at assessing the uncertainty associated with soil sampling in agricultural, semi-natural, urban and contaminated environments[6,7]. The "reference site" is located in the north eastern part of Italy (Pozzuolo del Friuli, Udine), in the research centre belonging to the Ente Regionale per lo Sviluppo Agricolo del Friuli Venezia Giulia (ERSA). The "reference site" is characterised in terms of the spatial variability of trace elements, and it is suitable for performing intercomparison exercises. The trace elements present at the reference site are of a combination of natural and anthropogenic origins.

Due to the limited extent of the reference sampling area and considering that collaborative field studies require considerable organisational efforts, only 10 ALMERA institutions were selected to participate in the sampling exercise.

Each participant laboratory was asked to apply its own soil sampling protocol (sampling strategy, sampling pattern, sampling design, sampling device, sampling depth, sampling techniques, etc.) and to collect not more than 15 soil samples.

For submission of the sampling protocols by the participants, a server has been prepared and established at the ICTP network, providing a web-based interface to the project database. Also all necessary information for the participants was offered via this system. The system was found to work in a satisfactory and robust manner, also as an example for possible arrangements for future exercises. In this frame:

- each participant laboratory provided a description of the sampling devices used and the sampling protocol applied during the intercomparison exercise (description of devices, number of samples, size of each sample, depth of sampling, etc.);
- each participant laboratory provided a description of its own methodology to prepare the test portion from each of the samples collected during the sampling exercise;
- each participant laboratory provided its own methodology to estimate the mean value of several analytes in a sampling exercise.

The participants' data will be evaluated according to IS0 13528[8]. The Atomic Energy Commission of Syria, following an agreed analytical protocol, carried out the determination of trace elements on the soil samples, collected during the soil sampling intercomparison exercise. The results of the intercomparison exercise are under evaluation by APAT (the Italian Environmental Protection Agency) and they will be presented and discussed during the fourth ALMERA coordination meeting, at the International Centre for Theoretical Physics (ICTP), Trieste (Italy) in November 2007.

2.3. Questionnaire survey on the analytical capabilities of laboratories of ALMERA members, research projects being implemented, etc.

A questionnaire form has been prepared to survey the analytical capabilities and research projects of laboratories of ALMERA members, and their needs for reference materials and other support. The questionnaire will be put on the ALMERA web site so that all ALMERA member laboratories can answer it on-line.

2.4 Development of analytical procedures of radionuclides in environmental samples

It is not easy to develop a recommended analytical procedure that all laboratories can accept, as a large number of various analytical procedures have been developed and published by many scientists and analytical procedures are being continually improved with upgrading of the instruments and materials for measurement of radionuclides. Understandably, most laboratories prefer to retain analytical procedures which they have used for long time and which have been proved to be reliable, unless there is a strong reason to make a change.

However, one of the most frequent requests which Member State laboratories make of the Chemistry Unit, Seibersdorf is that for recommended analytical procedures. Some laboratories that are undertaking a type of analysis which is new for them would like a reliable source of procedures, so that they do not have to "reinvent the wheel". They would like a readily-available reference which they can quote. If possible, they would like to have procedures available that are used at a large number of laboratories and therefore could be regarded as having been widely tested. In some cases, for example for sample collection, the final result may be strongly dependent on the method used, and use of a widely-accepted method may be important to assure comparability of results between different laboratories.

One important source of such information to date has been the IAEA Technical Reports Series No. 295[9]. However, this guidebook is now 17 years old and so the information has become somewhat dated. Some other publications have addressed various aspects, for example quantifying uncertainty in nuclear analytical measurements[10] but a consistent and more complete set of procedures would be desirable.

Therefore, since 2004 the IAEA's programme related to the terrestrial environment has included activities aimed towards the development of a set of procedures for determination of radionuclides in environmental samples. It is intended that as these are developed, they will be made them available to users, for example by publication in IAEA series and/or by placing them on the IAEA website. The latter option would allow those who are interested in a reliable source of procedures to search for them according to their requirements e.g. by analyte and/or sample type. It is not intended that the analytical procedures included should be regarded as "recommended" or "endorsed" by the IAEA for any particular purpose, nevertheless it is expected that the information will be a useful resource and starting point for analysts.

The approach being taken for development of specific procedures and methods is to first review the literature on a given topic, and then based on this review develop a method written in accordance with ISO guidelines (Fig.1). The activities mentioned above are meant to be of general use to a wide range of laboratories. In parallel, a set of procedures and methods needs to be developed for the ALMERA network of laboratories. The activity has started with a review of methods for determination of ^{210}Po in environmental samples[11].

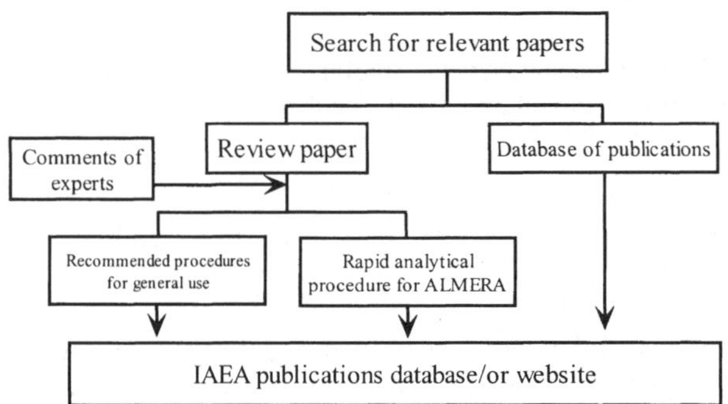

Figure 1 *A process for publicising of procedures for radionuclide determination in environmental samples*

Most research on ^{210}Po has focussed on its application as a tracer of environmental processes, and its impact on human health through radiation exposure[12, 13]. It has featured particularly prominently in studies of marine sedimentation processes, foodchains, atmospheric circulation and aerosol behaviour. Dosimetric studies have shown it to be the largest contributor of natural radiation dose to marine organisms[14] through ingestion, with resulting implications for human radiation exposure, particularly where seafoods are involved[15].

The determination of ^{210}Po is relatively straightforward, due to the ease of source preparation by spontaneous deposition onto metal surfaces and the uncomplicated alpha spectrum. Although several optimisation studies have been carried out, published source preparation methods remain remarkably diverse. For this review about 130 papers mainly focussed on analytical methodology of ^{210}Po were collected and critically examined. The literature surveyed included analysis of air, fresh water, rainwater, seawater, soil, sediment, coal, tobacco, phosphogypsum, foodstuffs, marine organisms, vegetation, human bone, and biota (Table 3).

Table 3 *Literature surveyed for preparation of ^{210}Po review paper*

Sample	No. of papers
Air	1
Drinking water	4
Ground water	3
Freshwater	6
Rainwater	4
Seawater	13
Soil	6
Sediment	9
Coal	1
Tobacco	8
Phosphogymsum	6
Food	9
Lake fish	1
Human body	12
Plant	5
Biota	13
Analysis methodology	29
Total	130

The volatility of Po was recognized as a problem in sample preparation early on, where losses begin at temperatures above 100°C, with 90% loss by 300°C[16]. This problem necessitates wet ashing techniques wherever possible in sample preparation.

Radiochemical yield tracers, ^{208}Po or ^{209}Po, are almost universally employed to allow correction for losses. The former has historically been the most widely used tracer, however ^{209}Po has advantages due to greater energy separation of the peaks, and a longer half-life resulting in a greater effective lifetime of the tracer solution.

The readiness with which Po auto-deposits onto metal surfaces results in employment of this in the final stages of most procedures to provide a source suitable for alpha-spectrometric analysis (Fig. 2). The time required for deposition may be minimised by adjusting the solution volume (preferably 20 mL or less), maximising the solution temperature (e.g. 80–90°C), and agitating by stirring, or by spinning the disc at, most commonly, pH 2. Given favourable conditions a deposition time of 1.5 or 2 hours should be sufficient. The interference of iron during electroplating is suppressed by addition of ascorbic acid[17] or hydroxylamine hydrochloride[18, 19] (to reduce Fe^{3+}). Citric acid may also be added to suppress effects of other ions present[20, 21].

A final analytical procedure for ^{210}Po based on the findings of the review is currently in preparation. The work on analytical procedures will be continued with a review of methods of determination of Pu isotopes by ICP-MS.

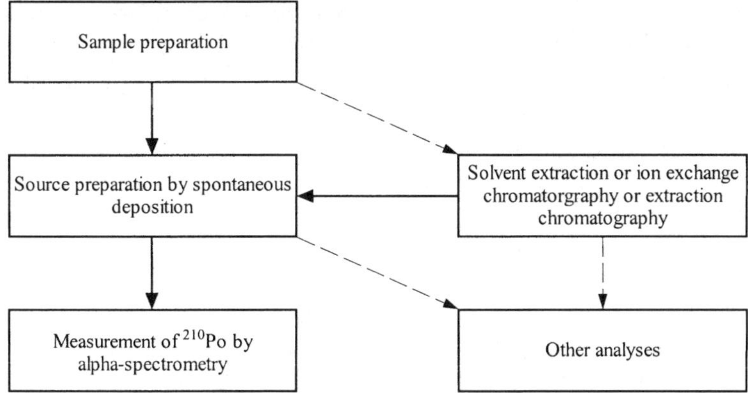

Figure 2 *A schematic diagram for analytical procedure of ^{210}Po*

References

1. Report of the first ALMERA workshop, IAEA/AL/113, 1997.
2. A. Shakhashiro, Z. Radecki, A. Trinkle, U. Sansone, T. Benesch, Final report on the proficiency test of measurement of environmental radioactivity (ALMERA) network, 2005, IAEA/AL/152.
3. Report on the second ALMERA network coordination meeting and the ALMERA soil sampling intercomparison exercise, 2006, IAEA/SIE/01, IAEA/AL/164.
4. Certification of reference materials-general and statistical principles, International Organization for Standardization (ISO), ISO GUIDE 35, 1989, Geneva, Switzerland.
5. General requirements for the competence of testing and calibration laboratories, International Organization for Standardization (ISO), ISO/IEC 17025, 1999, Geneva, Switzerland.
6. P. De Zorzi, M. Belli, S. Barbizzi, S. Menegon, A. Deluisa, A practical approach to assessment of sampling uncertainty, *Accreditation and Quality Assurance*, 2002, **7**, 182.
7. P. De Zorzi, S. Barbizzi, M. Belli, G. Ciceri, A. Fajgelj, D. Moore, U. Sansone, M. Van der Perk, Terminology in soil sampling (IUPAC recommendations, International Union of Pure and Applied Chemistry), Analytical Chemistry Division, 2005, 77, 827.
8. Statistical methods for use in proficiency testing by interlaboratory comparisons, International Organization for Standardization (ISO), ISO 13528, 2005, Geneva, Switzerland.
9. Measurement of radionuclides in food and the environment: A guidebook. Technical Reports Series No. 295, 1989, IAEA, Vienna.
10. Quantifying uncertainty in nuclear analytical measurements, 2004, IAEA-TECDOC-1401, IAEA, Vienna.
11. K.M. Matthews, C-K. Kim, P. Martin, Determination of ^{210}Po in Environmental Materials: A Review of Analytical Methodology. *Appl. Radiat. Isotopes*, 2007, **65,** 267.

12. N. Momoshima, Li-X. Song, S. Osaki, Y. Maeda, Biologically induced Po emission from fresh water. *J. Environ. Radioactivity*, 2002, **63**, 187.

13. P. Martin, B. Ryan, Natural-series radionuclides in traditional Aboriginal foods in tropical northern Australia: a review, *The Scientific World JOURNAL*, 2004, **4**, 77.

14. P. Stepnowski, B. Skwarzec, A comparison of ^{210}Po accumulation in molluscs from the southern Baltic, the coast of Spitsbergen and Sasek Wielki Lake in Poland. *J. Environ. Radioactivity*, 2000, **49**, 210.

15. United Nations. Sources and effects of ionizing radiation, United Nations Scientific Committee on the Effects of Atomic Radiation, 2000 Report to the General Assembly, with scientific annexes. United Nations, New York.

16. A. Martin, R.L. Blanchard, The thermal volatilisation of Caesium-137, Polonium-210 and Lead-210 from in vivo labelled samples, *Analyst*, 1969, **94**, 441.

17. R.L. Blanchard, Rapid determination of lead-210 and polonium-210 in environmental samples by deposition on nickel, *Anal. Chem.*, 1966, **38**, 189.

18. W.W. Flynn, The determination of low levels of polonium-210 in environmental materials, *Analytica Chimica Acta*, 1968, **43**, 221.

19. G.J. Ham, L.W. Ewers, R.F. Clayton, Improvements on lead-210 and polonium-210 determination in environmental materials, *J. Radioanal. Nucl. Chem.*, 1997, **226**, 61.

20. G. Jia, G. Torri, M. Petruzzi, Distribution coefficients of polonium between 5% TOPO in toluene and aqueous hydrochloric and nitric acids. Appl. Radiat. Isotopes, 2004, 61, 279.

21. T.M. Church, N. Hussain, T.G. Ferdelman, An efficient quantitative technique for the simultaneous analyses of radon daughters ^{210}Pb, ^{210}Bi and ^{210}Po, *Talanta*, 1994, **41**, 243.

Isotope Index

Subject Index